ROUTLEDGE LIBRARY
SOCIAL AND CULTURAL GEOGRAPHY

T0229624

Volume 7

GEOGRAPHY SINCE THE
SECOND WORLD WAR

GEOGRAPHY SINCE THE SECOND WORLD WAR

An International Survey

Edited by
R. J. JOHNSTON AND P. CLAVAL

Routledge
Taylor & Francis Group

LONDON AND NEW YORK

First published in 1984

This edition first published in 2014
by Routledge
2 Park Square, Milton Park, Abingdon, Oxfordshire OX14 4RN

and by Routledge
711 Third Avenue, New York, NY 10017

First issued in paperback 2016

Routledge is an imprint of the Taylor & Francis Group, an informa business

British Library Cataloguing in Publication Data
A catalogue record for this book is available from the British Library

ISBN: 978-0-415-83447-6 (Set)
ISBN 13: 978-1-138-97512-5 (pbk)
ISBN 13: 978-0-415-73326-7 (hbk)

Publisher's Note
The publisher has gone to great lengths to ensure the quality of this reprint but points out that some imperfections in the original copies may be apparent.

Disclaimer
The publisher has made every effort to trace copyright holders and would welcome correspondence from those they have been unable to trace.

Geography Since the Second World War

AN INTERNATIONAL SURVEY

Edited by
R.J. JOHNSTON and P. CLAVAL

CROOM HELM
London & Sydney

BARNES & NOBLE
Totowa, New Jersey

©1984 R.J. Johnston and P. Claval
Croom Helm Ltd, Provident House, Burrell Row,
Beckenham, Kent BR3 1AT
Croom Helm Australia Pty Ltd, First Floor,
139 King Street, Sydney, NSW 2001, Australia

British Library Cataloguing in Publication Data

Geography since the Second World War. —
    (Croom Helm series in geography and environment)
    1. Geography — History
    I. Johnston, R.J.   II. Claval, P.
    910   G62

    ISBN 0-7099-1411-3

First published in the USA 1984 by
Barnes & Noble Books, 81 Adams Drive,
Totowa, New Jersey, 07512

Library of Congress Cataloging in Publication Data
Main entry under title:

Geography since the Second World War.

    1. Geography. I. Johnston, R.J. (Ronald John)
II. Claval, Paul.
G99.G46   1984   910   84-2896
ISBN 0-389-20481-1

Printed and bound in Great Britain

CONTENTS

PREFACE

One would expect geographers, because of the nature of their discipline, to be well aware of intellectual trends within geography throughout the world. Unfortunately this is not so; in general, geographers are woefully ignorant of what their contemporaries in other countries are doing. Thus, this volume was conceived as an attempt to lay the foundations for greater international understanding among geographers.

This is not an encyclopaedia, treating 'Geography in . . . ' every country of the world in turn. The selection of which countries to stress is biased by editorial decisions (and even then, the anticipated chapter on Scandinavian geography failed to emerge). So, too, the selection of authors is biased, and they, in turn, were responsible for deciding which themes to stress. No blueprint was provided, and as editors we are delighted with the stimulating ways in which the contributors interpreted the general brief. We trust that readers will feel as better-informed as we now do.

In producing the book, we are indebted to the authors for agreeing to participate, for their good-natured responses to our pressure to meet deadlines, and for accepting our manipulation of their English-language presentations. We are indebted, too, to the editors of the series and to Peter Sowden of Croom Helm for their support of the project and their sympathetic response to 'deadline slippage'. Finally, we are grateful to Joan Dunn for an immense amount of secretarial work.

R.J. Johnston
P. Claval

1 INTRODUCTION: THE INTERNATIONAL STUDY OF THE HISTORY OF GEOGRAPHY

R.J. Johnston

The philosophy and history of their discipline are topics of increasing interest to geographers at the present time. There has long been a small body of geographers working on 'the history of geographical thought', but their concern until recently has been very largely with the period before 1945 and their approach has been chronological and descriptive rather than contextual and explanatory. The focus of attention now is the contemporary scene, especially the period since the end of the Second World War, with its rapid growth in higher education.

To date, research and writing on the contemporary period has been both tentative and partial. It has been tentative because workers have been uncertain as to the framework which they should place around their subject matter and as to the theoretical perspective that they should employ in their interpretations of recent events and debates. It has been partial because it has taken segments of the story — usually relating to particular countries and/or languages — and has ignored the events and debates elsewhere. With regard to the former, this essay is tentative too. But the aim — both of this contribution and of the book as a whole — is to improve the quality of the tentative theorising by attacking the second issue: that of partiality. The essays in this book treat the experience of geography since the Second World War in a variety of countries — though not, unfortunately, in all countries. These have been brought together to inform in two ways. Firstly, they inform readers of the developments in their discipline in other countries, developments of which they may well be unaware. Secondly, they inform workers in the field of the history of geography by providing further raw material on which their theories may be honed.

The present essay is both introduction and summary. It is introduction in that it provides the context for the study of this subject; it is summary in that it uses the material presented in the other essays to advance the contextual position.

The Context

Although work on the history of geography − other than chronology −
is in its infancy, already a number of contextual approaches have been
investigated. As yet, none has seemed particularly satisfactory, although
several have been popular − at least on a superficial level. This section
outlines very briefly the content of the apparently most attractive
approaches, before moving to a critique of the current situation.

Approaches to the Study of Disciplinary History

Undoubtedly the most frequently-employed approach has been that
developed by Thomas Kuhn (1962) in his pioneering book *The Structure of Scientific Revolutions*. In this, he proposed that any scientific
discipline's history is characterised by periods of normal science, with
occasional interruptions, or revolutionary periods. Under the conditions of normal science there is widespread consensus among practitioners relating to the nature of their discipline − what it knows, what
it needs to know, and how it sets about finding out. This consensus is
expressed in a *paradigm* − a (usually implicit) blueprint for the discipline. Because the contents of the paradigm are imperfect − some of
the knowledge is false, some of the methods used are inaccurate, some
of the questions asked are spurious, and so on − some of the work
undertaken fails, either absolutely or at least relative to expectations
(i.e. the predictions from 'known' theory). These failures create anom-
alies within the paradigm. They accumulate, and stimulate at least some
practitioners to seek alternatives, paradigms which are more successful.
When such apparently superior paradigms have been identified, they
must be promoted. A revolution is proposed, whereby all of the practitioners agree on the superiority of the new paradigm, and the old one is
overthrown.

Kuhn's model has a sociological element to it, for it involves
members of a discipline making what may amount to 'leaps of faith',
when they discard one paradigm in favour of another. This view was
criticised, for example by Popper (1972), who presented science in a
normative model of *conjectures and refutations*. According to this,
science is in a perpetual state of revolution as experiments and other
tests are designed to probe the validity of what we claim to know.
Knowledge is always open to query, to the possibility of being falsi-
fied, and scientists are trained to ask questions of assumed knowledge,
in an unbiased non-personal way, and to use methods the veracity of
which is clear. In this way, knowledge is built up by questioning and

testing, by discarding that which fails the tests, and by putting that which remains to further tests.

Somewhat akin to both of these approaches is that proposed by Lakatos (1978). With Popper, he accepts that the purpose of science is to put theories to the test, and, with Kuhn, he argues that most scientists will adhere to a particular theory or approach until it is successfully, to them, discredited. Unlike Kuhn, however, he does not believe that there are quick revolutions, and, unlike Popper, he does not believe that all scientific activity involves 'revolutions in permanence', or sequences of bold conjectures and critical experiments. Thus he formulated the concept of the *research programme*. This contains a hard core of irrefutable beliefs, and an outer shell in which the beliefs are put into practice. Within the shell, ideas are tried and tested, to advance the volume of knowledge attached to the core of beliefs; it is here that the conjectures are launched and the refutations are undertaken. But the core is unquestioned; it provides the driving force and the criteria for judgement. Several programmes may coexist in a discipline, competing for adherents as the shells are advanced through the acquisition of new knowledge.

These three approaches view science as an objective pursuit, with decisions being taken in dispassionate readings of the evidence. Their critics imply that such dispassionate readings are impossible, because they are undertaken by individuals, who are members of communities — in a general as well as a scientific sense. Thus other approaches are available that are based on the location of a discipline in its community context.

According to the *contextual approach*, the contents of a discipline reflect the demands made upon it by the society which nurtures it. Increasingly, work within a discipline is conducted by individuals employed by society, either in institutions of further and higher education or in research institutes. As paymaster, society has the ability to promote disciplines which it sees as profitable and to ignore, relatively if not absolutely, those of which it disapproves. Similarly, it can promote particular approaches within a discipline and downgrade others — in a variety of ways, including the resources that it makes available for research.

In its extreme form, this contextual approach suggests that society — via the state — dictates both the structure of scientific activity and the contents of the individual disciplines. Such a situation is unlikely. More likely is an interaction between society and the discipline, whereby members of the latter interpret what society requires and translate this into disciplinary action. In this way, the discipline, and the individuals, earn status and respect. Using these acquired attributes, they may then seek to promote their discipline to society, in particular via the institutions of the state, seeking to influence scientific and educational policy, to their own ends.

According to this view, then, the contents of a discipline reflect the *interpreters' views*. To some disciplinary historians – like political historians – this involves identifying the key individuals, those whose views have strongly influenced – if not determined – what has and has not been undertaken in their subject. Such individuals create *schools of thought*, with leaders (themselves), prophets and disciples; in some cases these may be national schools.

Although this last approach apparently has much to recommend it – for dominant individuals can be identified in the history of most disciplines – it fails to grapple with issues of change and size. With regard to the former, presumably the dominant individual can be associated with the core of beliefs in a Lakatosian research programme. Individuals can then choose between leaders and their schools. But, whereas in Lakatos's model the choice is an intellectual one, in the model based on the dominant leader it is a constrained one. In the development of a discipline the leader may exert substantial social control, over the allocation of employment and other resources, and favour adherents to the 'one faith' only. If this is so, then revolutions – or at least the advance of one view relative to others – are difficult to sustain, and may be generational, as one leader's influence wanes and another's waxes.

With regard to size, it is common experience that the larger the body the harder it is to hold it together, except in the face of some major threat. Thus the model based on key individuals is likely to break down when applied to modern disciplines, with many adherents spread worldwide. In these it is more likely to be the case that a relatively *anarchic* disciplinary organisation develops, comprising communities within communities with changing memberships. Each community may be linked to a particular research programme, and individuals may be able to move from one to the other with relative ease if they share the same basic orientation. Some may have periods of rapid growth, as they attract personnel and resources, whereas others may always be small and quiescent. Periods of relative success for particular communities may reflect an enhanced ability to obtain resources, perhaps as a reaction to a contextual need or as the result of the influence of a particular person.

In such an anarchic situation, a discipline proceeds on a lot of separate fronts. Its size precludes any centralised organisation of its research efforts, and its fragmentation is a natural consequence of the members' interaction needs and limits. Such a situation may be frowned on by those who remember an earlier era when smallness was associated with disciplinary coherence. But fragmentation has to be accepted as the price of growth, and growth must be promoted as a sign of political status and contextual relevance. However, the fragmentation must not be allowed to continue to its logical conclusion with the creation of new disciplines. Adherence — at least nominal — to the present discipline must be retained, because this is fundamental to the external projection of the discipline's image, especially necessary at times of societal questioning of the value of disciplines.

What, then, influences the operation of these communities within the disciplinary matrix? An approach relevant to the study of these is Foucault's concept of the *discourse*, a set of socially-defined rules governing and containing activity. Each community can be associated with its own rules and procedures, which provide the key to understanding its statements because they provide the definitions of its terms and the criteria with which judgements are made. The contents of the discourse are created by the members, in interactions among themselves and with the environment in which they are set (including other discourses). New members are socialised into the language and rules. The success of each discourse reflects its ability to attract and to hold members, which is a mark not only of its internal cohesion but also of its external image.

Parts of both the external image and the attractive force of a discourse reflect upon its spatial location. Knowledge is a universal commodity, especially published knowledge (Popper's World Three), but its availability is segmented according to its location. Undoubtedly the major cause of segmentation of the knowledge market is language; ideas are not readily accessible if they are expressed in an unknown or poorly-understood language (or jargon). Segmentation may also be ideologically or sociologically imposed — contact with certain ideas is prescribed for political ends, either those of the state and its institutions or those of key actors within a community. And the geographical factor of distance, too, may impede the spread of ideas. Thus discourses may be trapped within a variety of constraints that limit their spread.

The History of Geography

All of the approaches outlined in the previous section have been proposed as valid models for the study of the history of geography. All, however, have failed to convince when applied to the relevant material.

By far the most popular of the approaches during the last two decades has been Kuhn's paradigm model – introduced implicitly to the geographical literature by Burton (1963) and explicitly by Chorley and Haggett (1967). This model has been used to 'periodise' Anglo-American human geography since 1945 (see Johnston, 1979, 1983) and the concept of a revolution has been used in both a descriptive and a prescriptive sense (Harvey, 1973). In some analyses paradigms have been identified with particular individuals, thereby linking two of the approaches and implying that revolutions are created by particular scholars (Harvey and Holly, 1981), whereas in others the term has been used to describe any school of thought with a few adherents.

Criticisms of the paradigm model have been many. There is little doubt that it does not fit (Johnston, 1978, 1981, 1983) and that its concepts are largely irrelevant to academic developments within geography (Stoddart, 1981). By far the major element in the critique is the, readily-made, observation that, even if revolutions have been launched, they have not been complete. The apparently superseded approaches remain with some adherents, perhaps to reappear in a slightly altered form in later years. To account for this, some have proposed a modified paradigm model (Johnston, 1978, 1979), whereas others have been attracted to Lakatos's approach, which allows competing research programmes (Wheeler, 1982).

The biggest problem in the application of the paradigm model and those similar to it is that, not only does it apply to disciplines with a positivist base, but it also has its own in-built positivist methodology. According to the model's view of science (and Kuhn intended it as a model of the physical sciences only), progress is something that is measurable and there are precise criteria available for the verification or falsification of hypotheses. There may be debates over those criteria, but their applicability should not be in doubt. Thus objective judgements are possible, and progress can be identified as the accumulation of results.

It was this view of an academic discipline that characterised what became widely known as the 'quantitative revolution' in Anglo-American geography. The goal was the development of suites of laws that both predicted and explained human spatial behaviour and the spatial organisation of society. Progress involved greater methodological sophistication and the improved explanation that this brought.

This revolution was countered by arguments relating to the model of man that it implied, a model that debased human subjectivity and agency – except to represent it as a random element in a predictive equation (Claval, 1982; van der Laan and Piersma, 1982). This troubled those who identify human agency as the key element in the creation and recreation of a society and its artefacts, as well as those who identified the importance of structural elements directing society. Initially, it was thought that these alternatives could be identified as the precursors of revolutions. But the model does not fit. In the Kuhnian sense, a revolution involves a 'better' way of handling the same material, within a positivist philosophy. The human agency and structuralist approaches are based on different philosophies; they involve a switch in epistemological and ontological base, not just in methodology.

To replace the popular paradigm model, the research programme model offers some advantages but it too has an in-built concept of measurable progress and an implicit positivist philosophy. To an increasing number of geographers, these are irrelevant. Most of them are human geographers, and their changing orientations, in the 1980s in particular, furthered the break between human and physical geography. Until the 1950s/1960s, the two parts of the discipline were linked within the man-environment theme, although the interpretation varied: environment dominates people (environmental determinism); people dominate environments, to some extent (possibilism and probabilism); people and environments interact, to create landscapes (regions – usually regional geography had a strong trace of environmental determinism). In the 1950s there was a parting, in particular as physical geographers developed systematic specialisms. They were brought together again in the 1960s with a common methodological base and implicit positivist philosophy. Since the 1970s began, they have been growing apart again, because the common methodological base is no longer accepted by many. Thus one has a discipline, in some countries at least, comprising two very different parts: these would appear to have strong subject matter links; intellectually, they have diverged; academically, attempts are made to hold them together, for political reasons, and intellectual arguments are presented for that.

It may well be, then, that the paradigm model is relevant to the study of the history of physical geography. But what of human geography, part of which acknowledges a common philosophical base with physical geography, whereas the other part does not? For this, the contextual model has been promoted (e.g. Berdoulay, 1981; Capel, 1981). Interpreted as a direct social control, this model could be applied to physical geography as well. Its failing – for both – is that, although it may account for which disciplines receive resources and encouragement, because of perceived relevance, it does not account either for how the discipline acquired its particular image or for how intellectual developments took place within it. If any disciplinary autonomy is to be allowed, then the nature of decision-making – what to study, how to study it, who decides what is good or bad – is a separate issue from that of contextual constraints. For this, the key-individual model retains some attractions but, as argued earlier, such individuals are not necessarily apparent. Increasingly, it seems, the anarchic communities model is the most appropriate. The next section will explore why.

Geography and the History of Geography

The discussion so far has suggested that no single model of progress in a scientific discipline has been found satisfactory in its application to recent trends in geography. But most, I submit, have some validity and offer insights to what has been happening. In order to substantiate this claim, I look first at some recent developments in human geography, which provide valuable analogies for the study of disciplinary history.

The Theory of Structuration

A major debate within philosophy for many centuries, particularly with regard to the understanding of human behaviour, focuses on the relative merits of the polar opposites of determinism and voluntarism (Thrift, 1983). In *determinist* philosophies, the action of an individual is predictable, because it is governed by certain independent variables, either the physical environment (environmental determinism), transport-cost minimisation (locational analysis and regional science), rational choice in certain behavioural contexts (behavioural geography), or economic-structural forces (structuralist geography); the individual is a 'bearer of the message only'. In *voluntarist* philosophies, on the other hand, the individual is a free agent, creating a personal world and living within it.

To most workers, neither of these polar positions is particularly satisfactory. They see merits in both philosophies, and wish to combine elements from both. Human agency is presented as constrained but not determined, and decisions are individual but not free. Thus in geography, possibilism and probabilism were presented as more realistic interpretations of environmental influence, stochastic terms were introduced into deterministic equations to allow for 'irrational behaviour', and so on.

The most recent developments in this area in the English-speaking world relate to the ideas of Anthony Giddens concerning the concept of *structuration*. This is an attempt to situate human agency — individual and group decision-making — within a structural context, which comprises certain dominant forces within a society: such forces are social creations, however, and can be altered by social action (Gregory, 1981). According to Giddens, such forces are simultaneously constraining and enabling; they provide a context within which decisions are made.

Decision-making is a cumulative process; the results of previous decisions provide the context within which future decisions are made. And, as both Gregory (1978) and Thrift (1983) have argued, these contexts are spatially variable. (Hence Gregory's largely unappreciated call for a revised and revived regional geography.) Although, as writers such as Taylor (1982) have pointed out, the structural forces operate in a single (capitalist) world economy, many interpretations of these forces are possible. Once made and implemented, an interpretation constrains future interpretations — just as once a road network has been constructed it constrains future flows and development patterns (though, of course, the constraints can be over-ridden, by replacement of parts of the network).

The place-specific nature of many interpretations reflects a variety of factors. One is the partition of space into sovereign political entities, each of which presents its own interpretations of the structural forces and of reactions to them. (Frequently within any one country there are several competing interpretations, some of which have arisen as consequences of those with which they compete.) Another is the constraint of space-time. As Giddens (1981) has argued, developing the work of Hägerstrand and his disciples on time-geography, people operate within time-space prisms, and information is contained within them. (Its leakage is hindered by political boundaries and linguistic and other barriers, and by the well-known research finding that, although information may be widely available, its validity and value is often only recognised when reinforced through personal contact with a person already committed to a particular view.)

According to this — as yet far from fully worked out — structurationist perspective, regional geography is the study of different interpretations of the structural forces operating within the world economy. Each region is a context, comprising its physical environment, its economic, social and political structure, its culture, its built environment and its history (all of which are interrelated). It is in this context that decision-making continues, as geographical variations are created and recreated. Such geographical variations include variations in the discipline of geography itself,

A Structurationist Perspective on the History of Geography

The creation of a discipline of geography reflects the implementation of a latent need, one that is recognised and realised by one or more individuals within society. Thus Stoddart's (1981, p. 1) statement, that

> the history of geography is more than simply the chronological listing of the achievements of a few great scholars arrayed in national schools . . . both the ideas and the structure of the subject have developed in response to complex social, economic, ideological and intellectual stimuli

is somewhat misleading, for the great scholars (or, at least, the influential individuals) and the intellectual stimuli interact. Without individuals interpreting the stimuli of the times — successfully enough to win support — there would be no geography, and the nature of their interpretations has strongly influenced the content of geography ever since.

The history of geography is, of course, more than a charting of the beliefs of the influential individuals. Almost from the outset of the discipline as an academic subject there have been debates among leaders and their disciples — as in the Sauer/Berkeley school versus the Midwest school in pre-Second World War United States (Porter, 1978). But the views of those individuals have created the academic environments, creations which reflect both the local context (the cultural, political and other environments) and, in many cases, contacts (usually personal, often biased) with other contexts.

A successful innovator in such a situation probably creates an environment in which, over time, the intellectual power of academic leaders declines. This is undoubtedly the case in the United Kingdom, the United States and much of the rest of the English-speaking world today. The size of the geographical profession is undoubtedly a significant reason for this, particularly the rapid growth of that size in recent decades. Together these two factors have provided a context within which debate flourishes and pluralism of views prevails. The larger the disciplinary presence, the greater the probability of philosophical and methodological diversity (reflecting slight variations in individual interpretations of aspects of the discipline, variations which are transmitted generationally).

In countries where the geographical discipline is either smaller or more recent than is so in the English-speaking world, or both of these, the influence of key individuals is clearer. This is illustrated in several of the chapters in the present book. In many cases, the key individuals involved either come from countries where geography was already developed or were trained in such countries and returned home with a particular interpretation of the discipline. That interpretation, perhaps modified slightly to fit local circumstances, has become the local orthodoxy, and in some situations has been defended against attempts at altering the discipline. (See the chapters on Italy and the Iberian world: also Bosque-Sendra, Rodriguez and Santos, 1983.)

A simple relationship between the size of a discipline and the nature of its content should not be assumed, however. This is because the local context is both enabling and constraining, leading not only to different interpretations of the discipline by its practitioners (the separation of physical from human geography in the Netherlands is an excellent example of this) but also to different societal interpretations of what is expected of disciplines. Thus, in the 1960s in particular, geographers in the British Isles and North America were provided with a very permissive environment which spawned a large number of communities of workers — themselves generating others who reacted against established orthodoxies by creating others: as Johnston and Gregory note for the United Kingdom, this created disciplinary vitality at the cost of disciplinary coherence. In other countries this did not take place, either because of the continued power of individual geographers who defined the acceptable orthodoxy and, by their control of entry to the discipline, proscribed others, or by the institutional imposition of definitions of what a discipline's members are expected (even required) to do.

I do not, therefore, propose a continuum model, whereby the nature of the discipline of geography in all countries moves towards a UK/USA 'norm'. Developments in those two countries are likely to have some influence in most other countries — at various time-lags, depending on the nature of the international links and of the local environment. (It may be, of course, that the UK and the USA will again be influenced by developments elsewhere sometime in the future — as they were by work in France and Germany before the Second World War.) But the stress is on the voluntarism pole of the philosophical debate — geography in any country is what the geographers there (constrained and enabled by society as a whole) have made and are making it.

From the perspective of the countries in which the discipline is relatively large and the institutional and other constraints are relatively slight, we can identify a subject characterised by an anarchic community structure. Each community shares a perspective on the discipline which, at least implicitly, involves epistemological and ontological perspectives. Some of the communities operate in opposition to each other; most are separate but not independent, and it is possible for individuals to move from one to another, if not to work in more than one contemporaneously. Within some, the nature of progress is defined, and development akin to that described in Kuhn's model takes place (as, for example, in the quantitative analysis of spatial data). In others, the cumulative nature of the research is less clear – except in the quantitative increase of information available.

Summary

Ours is a fragmented world – politically, culturally, economically and socially. Academic and intellectual divisions follow this fragmentation, especially with disciplines such as geography in which the fragmented world is the focus of scholarly activity. Thus there is no one geography. There are many, reflecting varying interpretations of the nature of and need for such a discipline. Those interpretations involve not only the geographers themselves (who may differ considerably) but also the powerful elements within society that allow geography to exist. Thus in any place, geography is what geographers do – and that is a combination of what they want to do and what they are allowed to do.

Intellectual history reflects the interplay of ideas, in a societal context. Because the context varies in time and space, producing individuals who differ in their perspectives, so the history varies. In studying the history of geography, therefore, we can expect to identify no empirical generalisations or laws. All we can expect to find is a discipline structured, and continually being restructured, by the views of individuals operating in an environmental situation. To paraphrase a famous sentence:

Geographers make their own discipline, but they do not make it just as they please; they do not make it in circumstances chosen by themselves, but under circumstances directly found, given and transmitted from the past.

References

Berdoulay, V. (1981) 'The contextual approach' in D.R. Stoddart (ed.),
Geography, Ideology and Social Concern, Blackwell, Oxford, 8-16.
Bosque-Sendra, J., Rodriguez, V. and Santos, J.M. (1983) 'Quantitative
geography in Spain', *Progress in Human Geography*, 7, 370-85.
Burton, I. (1963) 'The quantitative revolution and theoretical geo-
graphy', *The Canadian Geographer*, 7, 151-62.
Capel, H. (1981) 'Institutionalization of geography and strategies of
change' in D.R. Stoddart (ed.), *Geography, Ideology and Social
Concern*, Blackwell, Oxford, 37-69.
Chorley, R.J. and Haggett, P. (eds.) (1967) *Models in Geography*,
Methuen, London.
Claval, P. (1982) *Models of Man in Geography*, Department of Geo-
graphy, Syracuse University, Discussion Paper 79.
Giddens, A. (1981) *A Critique of Contemporary Historical Mater-
ialism*, Macmillan, London.
Gregory, D. (1978) *Ideology, Science and Geography*, Hutchinson,
London.
—— (1981) 'Human agency and human geography', *Transactions,
Institute of British Geographers, NS6*, 1-18.
Harvey, D. (1973) *Social Justice and the City*, Edward Arnold, London.
Harvey, M.E. and Holly, B.P. (eds.) (1981) *Themes in Geographic
Thought*, Croom Helm, London.
Johnston, R.J. (1978) 'Paradigms and revolutions or evolution', *Pro-
gress in Human Geography*, 2, 189-206.
—— (1979) *Geography and Geographers* (first edition), Edward
Arnold, London.
—— (1981) 'Paradigms, revolutions, schools of thought and anarchy'
in B.W. Blouet (ed.), *The Origins of Academic Geography in the
United States*, Archon Books, Hamden, Conn., 303-18.
—— (1983) *Geography and Geographers* (second edition), Edward
Arnold, London.
Kuhn, T.S. (1962) *The Structure of Scientific Revolutions*, University
of Chicago Press, Chicago.
Lakatos, I. (1978) 'Falsification and the methodology of scientific
research programmes' in J. Worrall and G. Currie (eds.) *The Methodology of Scientific Research Programmes*, vol. I, Cambridge University Press, Cambridge, 8-101.
Popper, K.R. (1972) *Conjectures and Refutations* (4th ed.) Routledge
and Kegan Paul, London.

Porter, P.W. (1978) 'Geography as human ecology', *American Behavioral Scientist, 22*, 15-40.

Stoddart, D.R. (1981) 'Ideas and interpretation in the history of geography' in D.R. Stoddart (ed), *Geography, Ideology and Social Concern*, Blackwell, Oxford, 1-7.

Taylor, P.J. (1982) 'A materialist framework for political geography', *Transactions, Institute of British Geographers, NS7*, 15-34.

Thrift, N.J. (1983) 'On the determination of social action in space and time', *Society and Space: Environment and Planning D, 1*, 23-58.

van der Laan, L. and Piersma, A. (1982) 'The image of man: paradigmatic cornerstones in human geography', *Annals of the Association of American Geographers, 72*, 411-26.

Wheeler, P.B. (1982) 'Revolutions, research programmes, and human geography', *Area, 14*, 1-6.

2 FRANCE

P. Claval

Since the end of the Second World War French geography has experienced peaks and troughs, successful periods, periods of relative decline and periods of renovation. Between the two World Wars the French school had been in the forefront in the recognition of rich national diversity (Zeldin, 1973-7); in 1945 it was just stressing the urgency of the tasks which had to be faced by a France conquered and impoverished by half a century of Malthusianism and economic stagnation; it was moulding a perception of space which bore the complex diversity of the national spirit in mind. French geographers were read and appreciated in other social sciences (Duby and Lardreau, 1980): history, and particularly the *Annales* school, owed a great deal to it. Abroad, in the Anglo-American world and even more in the Latin world, French regional geography became a focus and served as a model.

At the beginning of the 1980s the discipline has lost some of its attraction for a society that takes most of its new ideas from economics, history and sociology. In the 1950s and 1960s, the period of tremendous growth, the sense of the diversity of the landscape and of its intrinsic values faded in the French society, but it re-emerged with the renewed interest in the rural past and traditional economies during the 1970s. Yet it was the ecologists and ethnologists who presented to young Frenchmen ideas they often took from the last 30 years of geographical work. In the French universities it became rarer for social scientists to consult geographical publications (Duby and Lardreau, 1980) and it subsequently became more and more difficult for geographers to make up for lost ground. On an international scale French geography had lost some of its distinction. This is partly explained by the decreasing use of the French language by British or American elites or by the elite in north European countries. The rest of the explanation lies in the reluctance to take up new ideas at the right moment.

Can we therefore talk about a decline? Not on every front: the increase has been considerable with regard to people, equipment and publications, linked as it was to the development of higher education,

15

service activities, administration or planning. Other disciplines have expanded more quickly, but there are many more teachers and researchers in geography than 30 years ago. Although the expansion was slow until 1955, the pace quickened between 1960 and 1972, which was the end of the period of rapid growth. The discipline did not have such a glowing reputation, but the printing of geographical material was actively carried on and a remarkable number of books was printed, above all volumes containing depictions of rural life and regional landscapes. French geography was less universally acknowledged by other disciplines, but a considerable revival of interest had taken place during the last ten years and many historians, economists, sociologists and political scientists have begun to read geographical publications again. A new interest has also been shown abroad, including certain fashions, like the one for urban studies by the French Marxist school in the 1970s.

Changes of direction become apparent when the evolution is examined more closely. The discipline's first troubles coincided so precisely with the period of rapid expansion that the 1950s and 1960s are marked both with material growth and by a weakening of most of the traditional positions.

The turning-point in the 1970s is more pronounced than in other countries; this is due to the regression caused by May 1968 which had a profoundly disturbing effect on the university system, but it is also due to the decline in the numbers of geography students between 1969 and 1974. The university crisis which affected geography in France thus dates from 1972 or 1973, five or six years before the crisis which hit the other social sciences or which affected materially our discipline in other industrial countries. This has made the task of recovery even more arduous.

The Material Conditions

I have described elsewhere (Claval, 1976) the structure of teaching and research in France, the role of the universities (which today include 606 teachers compared to 544 in 1972 and 71 in 1955) and the role of the ORSTOM (*Office de la Recherche Scientifique des Territoires d'Outre-Mer* – Scientific Research Office for Overseas Countries) and CNRS (*Centre National de la Recherche Scientifique* – National Centre for Scientific Research). Over the last ten years there has been no significant improvement in the equipment provided for researchers except in the sphere of information processing, something with which universities and geography departments were poorly provided. The construction of new buildings, which proceeded so rapidly between 1955 and 1973, has been brought to an almost complete standstill.

French geographers are thus faced with a situation which is often very grave. The people who teach in the universities are well paid and most of them have a secure job, but their premises are unsatisfactory (especially in Paris), the facilities provided by the university laboratories are limited, and the money granted by the CNRS is precariously unreliable. It thus became necessary to look for new sources of finance: contracts have been signed with the *Ministère de l'Equipement* (Ministry of Roads, Housing and Urbanism), the *Ministère de la Coopération* (Ministry for Co-operation with Third World Countries) and the DGRST (*Direction Générale de la Recherche Scientifique et Technique* – Direction of Scientific and Technical Research). Geographers are not always successful in breaking into this field compared to their principal rivals – sociologists and economists: the very structure of geography departments impedes the formation of big teams, a necessity for the successful attainment of research contracts.

A large amount of systematic research, demanding heavy injections of capital, escapes the universities and CNRS which do not concern themselves with this. As far as remote-sensing is concerned, for example, the most efficient teams belong to the *Centre National d'Études Spatiales* in Toulouse or to the *École des Mines* in Paris (although the laboratory of *École Normale Supérieure* is by no means negligible in this respect). The IGN (*Institut Géographique National* – French Ordnance Survey) and the statistical department at INSEE (*Institut National de la Statistique et des Études Economiques* – National Office for Statistics and Economic Studies) are mainly responsible for the development of automated cartography.

Colleagues of other disciplines cover certain fields which elsewhere are open to geographers: French regional science is still mainly undertaken by economists; historical geography attracts as many historians as geographers; environmental problems have offered more opportunities to ecologists than to geographers; the rediscovery of the rural past is primarily due to the work of ethnologists and folklorists. In return, geographers are inclined to be given a free hand with subjects that are either little taught or badly taught elsewhere – oceanography, climatology, meteorology, demography.

The number of students, which quadrupled in the 1960s, has almost returned to its original level: the trough of the wave came about 1975.

The decline was linked to the change in the entrance requirements for education in the social sciences (geography ceased to be the only discipline open to people who had not studied Latin) and was due to the dislike for a subject in which it was difficult to find employment. Since 1975 prospects have not improved in this respect: teaching has traditionally always been the most important opening, but this has been cut due to the population decline and the new method of recruiting teachers in the *premier cycle* (the eleven to fifteen-year-olds' part of the French school system). New job opportunities have been created in planning, but they are scarce and the competition is severe; geographers must prove themselves against economists, sociologists, graduates from the *Instituts d'Études Politiques* (Institutes of Political Sciences) and increasingly against engineers who have studied for *doctorats de 3° cycle* (Ph.D.) in economics or planning. Often the only sector in which geographers predominate is physical planning, in the elaboration of the *Plans d'occupation des sols* (Zoning Regulations) laid down by the 1967 *Loi d'orientation de l'urbanisme* (the basis of modern town planning in France). Without the openings created by the implementation of this new legislation, the prospects would have been even more grim.

French geographers are probably most favoured when it comes to the publication of their research results. It is customary to give priority, at almost every level, to the dedication of a large amount of the resources to the printing of journals and monographs. Most provincial departments have launched collections of works. The number of geographical journals has not declined, despite the sharp rise in printing costs; this is due to the grants from the CNRS and to the energy of the people who manage and encourage their production. It is regrettable that most of these publications are regional studies which means that important sectors of a more general geography are insufficiently covered.

Many different associations working closely with each other define the structure of French geography at a national level: the Association of French Geographers, the National Geographical Committee, groups of professors and lecturers. Since 1963 annual geographical days have provided a useful setting for people to meet; the numbers present reached a maximum at the beginning of the 1970s when about 500 people attended these reunions; now there are hardly ever more than 350 people present. This reflects a despondency on the part of the profession, but above all it is due to the transfer of work to specialised sections such as rural geography (which served as a model, under the chairwomanship of Jacqueline Bonnamour), urban geography, industrial geography and tourism. Each branch organises its own meetings, excursions and seminars. Researchers in the same specialism have the chance to exchange ideas on these occasions.

Since 1970 sociological organisation has partly escaped its traditional structure. New points of contact have been created without the participation of the most eminent professors. A number of them practise their skills in the provinces, such as the Dupont group which usually meets at Avignon (hence the name) and brings together an important group of geographers from the east and the south-east of France to introduce them to quantitative methods and to develop theoretical geography. The *Géopoint* colloquium, held every two years, unites researchers attracted by new thoughts on the basic principles of the discipline or about one or other of its aspects. In Paris *L'Espace géographique* has played a similar role in organizing seminars, colloquiums and meetings for foreign geographers or for researchers in other social disciplines.

French geography has been developing in a difficult context over the last ten years, although it has escaped the brutal reductions in staff which have hit those countries where the crisis struck later but where job prospects were less secure. The economic climate was not equally favourable to all branches of research. It did, however, allow interesting work to continue in a number of areas; despite these adverse conditions French geography has managed to make up for some of the backwardness engendered by the previous 20 years of ease.

The Initial Situation, 1945-50

Applying to geography a formula developed for history by Marc Bloch, T.W. Freeman has spoken of the 'geographer's craft'. The expression is very helpful for the understanding of French geography in the 1940s. It was not concerned with a subject defined by a body of structural concepts, by knowledge which is presented in such a way that it fits into a framework. Geography was not thought of as a general science but as a subject that one learned through practice, or rather, in the field. In order to become a geographer it was necessary to learn to map-read, to pick out significant characteristics of a landscape — its relief and agrarian structure — and to interview people about their way of life and about the way they work in and understand their environment.

Practised in this way, geography was built around a core which assured its coherence: it began by noting density and its goal was to describe the regional organisation of space, to explore its origins, and to show its significance to the groups involved (Figure 2.1). This way of

Figure 2.1

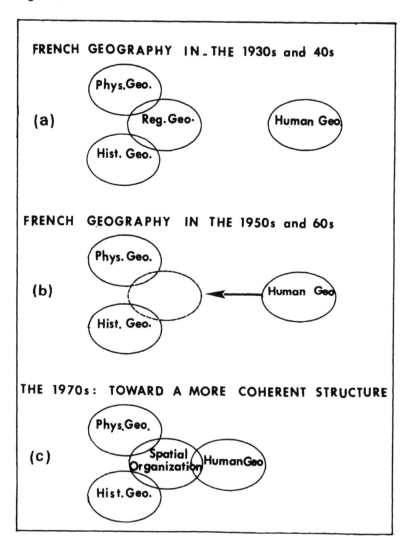

carrying out research entailed certain methods: morphological analysis, which was necessary for an understanding of both relief patterns and forms of rural life; archeological surveys of the characteristics of the landscape, which allowed one to interpret the history and formation of these elements by observing their present conditions; interviewing people with access to special information, such as influential people in the area or old farmers, the only people able to show how the diverse communities have amalgamated and how each basic group — family or rural enterprise — fits into the general scheme.

If geography is to function in this manner, it requires the fusion of physical geography and historical geography to the solid core of regional geography (Figure 2.1a). The study of relief features allowed one to understand the environment in which people had settled, and its origins, and to measure the constraints which it imposed on people. Climatology thus proved very helpful, especially for those areas that lacked water, and those areas that were subject to unpredictable hazards and the risks that these bring for the environment and for man.

An historical mode of analysis was indispensable for an explanation of origins much more than for a functionalist approach, but there were no specialized historical geographers. French geographers were, by training, both historians and geographers and vice versa. Those students whose main interest was in the past specialised in history; those whose interest centred on the present became geographers. A tendency therefore developed whereby the reconstruction of past geographies was left to historians. What was demanded from history was the illumination of the emergence of the significant characteristics of the present world: this method was used by Pierre Deffontaines (1932) in his study of *Les Hommes et leurs travaux dans les pays de la Moyenne Garonne*, in which he looks back only as far as is necessary for an understanding of present life — to the wave of medieval occupation for agrarian landscapes and forested areas, to the sixteenth century for trade and to the eighteenth century for manufactures.

Bearing this viewpoint in mind, can a role be found for a general human geography apart from a presentation of the vocabulary used in regional description? What existed as general research (in the economic and particularly in the political fields) did not bear much relation to the central core of the discipline (Figure 2.1a): this is clearly seen in the contempt displayed by university researchers faced with economic geography, a supplementary area useful for professional teaching but hardly worthy of sustaining real work. Political geography was undoubtedly more favourably received, but it passed for an applied field of study rather than for an essential branch of geographical knowledge.

The view of the discipline which I have summarised was formulated at the beginning of this century but was hardly developed in the 1920s and 1930s. Progress in research had been shown by the interest which physical geography sustained, by the spate of regional inquiries, and by the increasingly systematic use of archival documents to elucidate the origins of contemporary life. If there was a difference between the geography of the 1940s and the geography of the 1920s and 1930s, this was where one needed to look. Following Gaston Roupnel's intuitions (1932), Marc Bloch's synthesis (1931), and Roger Dion's work (1934), the history of agrarian landscapes had become a major field of inquiry, people delved further and further into the past and some people began to think that the geographer's craft was to elucidate the origins of the structures of land use.

The evolution of a scientific discipline stems from the general history of the country in which it develops and from the phases of its demographic development. Traditional French geography developed from the work of a group of worthy researchers, most of whom were born between 1870 and 1885: de Martonne, H. Baulig, J. Brunhes, A. Demangeon, R. Blanchard, M. Sorre. They were very active in the 1920s and most of them remained active in the 1930s. By the end of the Second World War, most of them had disappeared; Max Sorre was the only one to have written any major works during the 1940s and 1950s. The following half-generation, those born between 1885 and 1900, was much more poorly represented: decimated by the First World War or desparate after years of national service, its representatives are rare: André Cholley and Georges Chabot, but neither of these possessed very wide-ranging interests. For those who were born later, between 1900 and 1915, the possibilities of a university career were greater – P. Birot, A. Cailleux, J. Dresch, A. Guilcher, P. Pédelaborde, P. Gourou, R. Dion, M. Le Lannou, P. George and E. Juillard, for example – but their work did not have any impact on ideas and methodology until just after the Second World War. French geography lacked first-rate men in the 1930s and 1940s and those who rose did not have the authority to change the direction of the subject or to restructure it in any significant way.

Crisis, 1950-68

The Decline of Regional Geography

The 1950s and 1960s witnessed a crisis in French geography. It was a long time before French geographers became aware of this: did they not consciously continue to apply the principles that were perfected during the preceeding period, those that had made the regional school famous abroad? This was certainly the case, but the crisis reflected a change in modern French life rather than a despondency within the discipline.

The regional methodology was easier to apply in those areas in which there were still simple activities that were expressed directly in the landscape than in those areas in which the means of production were more developed, relations more complex and the correspondence between form and function less obvious: this methodology had essentially been used in rural France. The rural areas changed very little at the beginning of the century; isolated from the outside by Méline's protectionism, agriculture clung obstinately to many of its traditional structures, which meant that inquiry was necessarily drawn to the distant past.

After 1945 progress in France accelerated. Rural areas were modernised, became severed from their past and were rapidly emptied. An historical approach was no longer of much use in explaining these changes: regional geography was powerless faced with a world that could no longer be explained by its past but which was illuminated by its present functioning.

Tropical Geography

The regional paradigm remained valuable only with regard to developing countries, therefore. Paradoxically, at a time when French geography was faced with a crisis at home, it produced outstanding works in the tropical world. Generalizations made by economists and sociologists are not very helpful when one is concerned with small and divided social groups singled out by the weight of diverse traditional cultures and restrictive environments, in understanding resistance to progress and conditions of modernisation which are not too traumatic to undergo. In this case there can be no substitute for geographical or ethnographical monographs. Under Gourou's impetus (1947), tropical geography continued to develop actively and to take advantage of the close link between regional and physical research which very quickly fostered a strong interest in the recent development of ecology. It also drew upon the results of anthropological research in Africa and in Southern Asia and upon the results of sociology in South America: without this it would be impossible to understand the originality of the works of P. Monbeig (1952), G. Sautter (1966) and P. Pélissier (1966) to name but a few. Regarding J. Gallais (1967), the discovery of the living experience and of the problems of perception have breathed new life into this research.

The favourable position that it holds with respect to public aid shows the strength of tropical geography. Most of the money comes from ORSTOM, which supports teams in most of the countries of the former French colonial empire; it has also benefited from the creation, under the leadership of Guy Lasserre, of an influential laboratory for tropical geography at Bordeaux, the CEGET (*Centre de Géographie Tropicale* – Centre for Tropical Geography).

Historical Geography

The fading-out of regional geography in France was at first characterised by the increasing independence of physical geography and historical geography. They both benefited from this but not to the same extent (Figure 2.1b).

The popularity of historical studies reached a peak in the early 1950s when most regional geographical theses stressed the origin of agrarian landscapes (for instance, Juillard, 1953; Lebeau, 1955; Chevalier, 1956; Flatrès, 1957; a synthesis is Meynier, 1958). The results that were compiled received quick recognition and even today the colloquiums on the rural landscapes still follow the same methods that were outlined then. This success should not be exaggerated: for most French geographers what still matters is the explanation of the present; a study bearing less and less relation to contemporary concerns will only attract a minority of colleagues. The competition from historians is another concern in this field of study. Since Lucien Febvre, the revival of French history has been greatly indebted to the methods and interest which have been inspired by geography. The geo-historical literature, written in a manner similar to Fernand Braudel's (1949), is scattered with ideas from Vidal de la Blache and Jules Sion; theses on regional history devote more and more space to the reconstruction of landscapes and agrarian economies of the past. Is it really surprising then that the most stimulating work carried out by Roger Dion, published over twenty years ago, *L'histoire de la vigne et du vin en France* (1959), did not have any immediate successors?

Out of the historical field grew, nevertheless, some fascinating studies in cultural geography (de Planhol, 1958, 1968), but with no immediate posterity too.

Physical Geography

Physical geography has enjoyed a more stable success. For the student joining the discipline in the 1950s, it represented the most structured sector of geography, the only one united by a coherent theoretical basis. The attraction was even greater as French morphologists were on the brink of a renovatory movement: the interest in climatic-control, which stresses form and escapes Davis's normal cycle of erosion, was already apparent in the work of de Martonne in the 1930s, and in the thesis of J. Dresch (1941). P. Birot (1955, 1959), J. Tricart and A. Cailleux (1953-65), and M. Derruan (1956) opened a completely new world to research. In exchange for the rather abstract methodology of Davisian geomorphology, approaches were substituted which discussed deposits left by former erosion processes; when they were analysed in the laboratory they illuminated the conditions necessary for the origins of past forms. It became more difficult to outline grand historical syntheses, but their results were based on sounder methods. The Mediterranean world, North Africa and the Sahara, Western and Central Africa and Madagascar, which were easily accessible to French geographers throughout this period, facilitated the rapid development of a new climatic and zonal geomorphology.

The bond between botanists and soil scientists became stronger at this time, and this is expressed by G. Rougerie's African work (1960) and in the work carried out in Europe by G. Bertrand (1968). Biogeography, which for a long time had been left to botanists such as H. Gaussen and P. Rey, began to attract geographers (P. Birot, 1965; G. Bertrand, 1968). Oceanography, as shown by A. Guilcher (1965), attracted a large number of young investigators. The climatology practised between the wars, in particular by de Martonne, had remained classificatory and made scant use of the modern ideas of dynamic meteorology. By exploiting these ideas, P. Pédelaborde (1957) was able to outline new avenues of research in the 1950s.

There is, therefore, nothing surprising about the favour accorded to physical geography in these two decades. The CNRS was sensitive to this issue and set up a large laboratory at Caen, with A. Journaux as director. It supports university research and provides equipment which could not be dispersed without incurring tremendous wastage; French geomorphology commanded experimental facilities which were lacking until then.

The success of physical geography was not untainted. In breaking the bonds which united it with regional geography it has undoubtedly been able to develop more easily according to its own dictates — but this has led to increasing specialisation and means increasingly that a geographer must be a geologist, mineralogist, chemist, botanist, meteorologist and oceanographer at the same time. This attracted students who wanted to be naturalists rather than geographers and it discouraged people interested in cultural phenomena — especially since the French university structure, even after 1968, does not make it easy for one to follow two specialised courses which straddle the earth sciences or physical sciences, and the social sciences.

Throughout this period physical geography loosened its ties with human geography, so that what was common to both fields — the ecological approach, the analysis of environmental conditions — was neglected. G. Rougerie (1969) expressed an interest in this approach, but few French geographers began to study environmental problems or natural hazards and very few began to analyse human-induced erosion. The progress made in physical geography did not play much part in the restructuring of the aggregate field of the discipline. Research which demands a considerable amount of equipment and huge investment is not easily adapted to a state of administrative disorder: directions followed in the 1950s and 1960s exposed physical geography to much disappointment when the old university system broke down following the events of 1968. Many people were so affected by this that they tried to escape from geography departments to join scientific sections that were less shaken by the crisis.

Rethinking Regional Geography

Does this mean that specialists in human geography did nothing when faced with the crisis which affected regional geography? As early as the early 1940s A. Cholley (1942) had considered it necessary to rethink the latter, to place it in a firmer conceptual setting, and to free it from the muddle it had been in until then. Such a reorientation entailed more than just a redefinition of vocabulary — it necessitated original thought on what unites men and structures space: this would only be possible by rethinking the fundamentals of human geography as a whole. Since the early 1950s researchers have tended to steer the discipline towards a systematic study of the factors which organise space — but the approach was difficult and no firm conclusions were reached in this period.

At the end of the 1940s M. Le Lannou (1949) had proposed, by borrowing some of Vidal de la Blache's intuitions, to rebuild geography around the central idea of man as a dweller. The idea was not perfect as it hardly permitted the forging of a link with the powerful new developments of sociology and economics, but it had the advantage of introducing concern for subjective factors into human geography, something significant for the people being studied. Eric Dardel

(1952), a philosopher, demonstrated the coherence of such a scheme in the early 1950s, but the idea was not taken up by French geographers at that time.

Étienne Juillard reused the theme developed by Cholley to make a thorough study of the framework of regional geography (1962). He went beyond the elaboration of a new set of definitions: he retraced the evolution of ideas and stressed the role of the economic and social forces which structure space, although he did not make a systematic study of these forces: his efforts are more useful for the explanation of former patterns than for explaining forces at work in the contemporary world. The interest that he increasingly displayed for historical geography and for the reconstruction of past regional structures bore witness to this fact.

Max. Sorre proposed a more extensive reorganisation than the one devised by his young colleagues. He set about redefining human geography (1943-52) and its relationship with sociology (1957): many people were quick to understand the implication of his thoughts, particularly Paul-Henri Chombart de Lauwe (1952) who used them in his studies on Paris. In the 1950s sociology was, alas, lacking any satisfactory framework that could be directly applied to the analysis of spatial patterns: geography could help sociology but had only a slight theoretical body of interpretation to borrow from it.

Demography had certainly more notions and material to offer to geography and it was widely used during his period (George, 1951).

Economic Geography and Marxist Influences

Geographers seeking to construct a new geography turned next to the economic sciences. There was important progress going on in spatial economics at that time in France (Perroux, 1950; Ponsard, 1955; Boudeville, 1961), but geographers preferred to use more empirical approaches (Labasse, 1955). Most of them followed in Pierre George's footsteps in not opting for classical economics (1945). This was readily considered suspicious by an intellectual class stamped by Marxism – and French geography at the beginning of the 1950s was probably the discipline in which the communists were the most influential. Most young researchers thus inclined to this direction. What could they teach then: to be interested in the modern forms of the economy; to study the structuring of business concerns; to trace the origins of the capital which these concerns used? They were also taught to investigate the relationship between the town and the countryside (Rochefort, 1960; Dugrand, 1963; Brunet, 1965) – the French 'intelligentsia' had discovered this in the famous pages of the *German Ideology*, the first translation of which had just appeared.

The interest in the Third World did not begin with the Marxists, but they used their own theories on the Asian mode of production or Lenin's arguments on imperialism to explain inequality (Lacoste, 1965). In developed countries attention focused on the level and patterns of consumption which allowed an outline of classes to be drawn. These were new concerns: geography lost its retrospective aroma and turned towards the most significant developments of the age.

Does a Marxist framework lead any further? The answer is no and this was soon apparent. In its most popular form in the 1950s and 1960s, the key notions of a Marxist interpretation were the means of production and productive forces: space played no part in the essential analysis. Geography only pays attention to the succession of social formations in every region: it reveals the diversity of the world but does not explain it. A Marxist human geography cannot be successful if it starts from these premises. This began to dawn on researchers in the 1960s. If you add to this the disillusionment brought about by the revelation of the horrors of Stalinism, of the crisis of Budapest and the recognition of the backwardness of Russian society, you can understand the ebb of the Marxist wave. But the experience is characterised in the long term by graver repercussions. The other social sciences, less influenced by Marxism, ceased to take geography seriously; all that one found in it was conventional analysis which shed no light in the explanation of the origins of spatial patterns. The disinclination to help capitalist society to overcome its contradictions led some people (George *et al.*, 1964) to condemn the first efforts of applied geography. In fact, despite George's criticism, applied geography began to develop (Phlipponeau, 1960; Labasse, 1966) and the publication of regional atlases launched by Jacqueline Beaujeu-Garnier was welcomed by planners.

The proclaimed hostility of Marxists to anything that was connected with the United States led to an ignorance of Anglo-Saxon geography, new theoretical approaches and the quantitative revolution.

International Contacts and Institutional Change

However, one must not exaggerate the responsibility of the Marxists in bringing about this situation: traditional French geography was more a craft than a scientific discipline; it hardly acknowledged methodological debate and foreign works were consulted only to glean their results.

Was it right occasionally to model ourselves on others? Yes, undoubtedly, but the example to follow was the German one. It was the German language which was recommended to students, and many colleagues born around 1900 spoke German. After the 1930s the German school no longer played a key role in the discovery of new ideas and France lost contact with the cradle of innovation.

The new Anglo-Saxon geography had taken its theoretical models from the economic sciences and it had drawn its methodological inspiration from the English tradition of statistical analysis and from the recent developments of operations research. This field was unexplored by most French geographers who had difficulty reading English, had had a literary training, and were discouraged by an antiquated way of teaching economics at the university, so they gave spatial economists a free hand in this area. Some people built up an economic geography capable of explaining spatial organisation, by using the analysis of economic flows and distance-decay functions. They emphasised the role of transportation costs and of the costs of communicating information (Claval, 1963, 1968a, 1968b). They already understood the limits of this economic model and the necessity to include social science and politics in their explanations (Claval, 1967).

During the 1950s and 1960s a new interest in methodology (Brunet, 1968) and in the history of geography (Claval, 1964) began to develop. P. Pinchemel organised a group of young geographers focusing on this scheme.

Although the 1950s and 1960s were years of intense work and thought, the results do not measure up to the effort expended upon them. Geography had developed in a material sense but without anyone being able to define its boundaries and to understand completely the uneasiness which oppressed it. This is what lent significance to the uprising of 1968. The questioning of the professors' authority went beyond a straightforward criticism of the university hierarchy and harshly revealed the identity crisis which had been brewing and the need for a deep analysis of the situation. Everyone was implicated in the task of reconstruction. This took place in difficult circumstances: the disorganisation of the university administration, political and union restlessness, and the fact that money had been cut increasingly since 1972 — but it led to a change that we must now appraise.

Restructuring Since 1968

The Search for a Focus

The 1968 crisis left the universities in a state of chaos. The power of traditional leaders was cut considerably and the young lacked either anyone with an understanding of the new political directions or anyone who was able to suggest a new way of organising the discipline. This explains the confused development, the numerous directions and the lack of a universal aim. Co-ordination no longer came from traditional channels — those of the university hierarchy. A few years were needed in order for other structures to be developed — special sections of the National Committee (the French section of the International Geographical Union — IGU) or informal groups. Now some dominant new trends are appearing. Geography is becoming very receptive to outside influences. The present generation is willing to read English and speak it to varying degrees of fluency. Ideas expressed in major Anglo-Saxon periodicals are beginning to reach them, whilst many colleagues have the opportunity to experience a more direct contact with Anglo-Saxon geography through periods of teaching in Quebec. It is in this way that quantitative methods have penetrated France.

The aspirations of young French geographers are in many respects similar to those that one finds in advanced industrial countries: the need to find employment and the worry about being useful were already beginning to be felt in the previous period (this was shown by the unpopularity of retrospective studies). What is newer is the interest in other models of man and society. French geographers have also become aware of environmental problems, the pressure on the environment and the quality of life (Rougerie, 1975). The French intellectual situation is nevertheless so unique that influences and shared interests are not enough to bridge the gap between the evolution of French geography and geography abroad.

Radical Anglo-Saxon geography developed as a critique of the reductionism of the economic models which prevailed throughout the preceding years and in response to the limitations of positivism implicit in the systematic and preferential use of quantitative methods. Economic models, as has just been shown, had only a limited success in France and their inadequacies were in fact highlighted by the people that first began to use them. The fact that most French geographers have had a historical training has always made them sceptical of the allure of an ahistorical positivism; they know that they are dealing with specific societies which are continually changing, and the hope of seeing laws at work in these societies, laws which are independent of time and place, has never been shared by the majority of them. The rigid geometry employed by certain Anglo-Saxon theoreticians in the 1960s is not fitting, for them, and has never fitted the actual arrangement of space.

The criticism of a positive geography in Anglo-Saxon countries has been launched in the name of humanism or Marxism, which thus seem naturally to succeed an outmoded positivism. French geographers have never forgotten the lessons of traditional humanism; dressing it up as modern phenomenology does not change much, and in any case Eric Dardel used this method in the 1950s. As for Marxism, at least in the forms that it took in France in the 1950s and 1960s, everyone is assessing its weaknesses and limitations and many people share the view that it has hindered development much more than it has helped it (Claval, 1977). At the beginning of the 1970s, when French sociologists and urban economists (Castells, 1972; Lipietz, 1977; Lojkine, 1977) turned Marxist after Henri Lefebvre's example (1974), geographers followed their work with a good deal of scepticism. Should one be interested in problems of rent and in the land market? These are surely some of the key factors that structure space in our societies. The transition from this position to following in the footsteps of Castells or Lipietz is a gap that most people would refuse to cross. Can one talk about the 'production' of space? Does it not sound queer? The expression aims at stressing the historicity of everything that surrounds us, including the most unrefined areas, but it does not change the way that we proceed at the moment and seems a pedantic way of dealing with a simple idea.

There are many people who have left the Communist Party but who emotionally and intellectually remain Marxists. They know that the dogma propagated in the 1950s and 1960s cannot replace geography, but the hope that Marxism will take a new course once it has rid itself of its dogma has not been lost. This explains their interest in new schemes, for example urban rent and the relations between centre and periphery, and their efforts to use them, but also the care with which the experiments are undertaken. Marxism is on the defensive in France when it comes to geographical explanations.

Most geographers are deeply aware of a crisis to which there is no easy solution. It is imperative to break new ground, to use quantitative methods and theoretical models developed in economics, but everyone knows beforehand that this will not solve all the problems. The search for renovatory principles is thus almost universal (Beaujeu-Garnier, 1971; Dollfus, 1971; Reynaud, 1974; Lacoste, 1976; Isnard,

1978; Isnard, Racine and Reymond, 1981; Bailly and Beguin, 1982). It has also strengthened the interest for the origins of French geography (Broc, 1975; Berdoulay, 1981).

There are no longer any fields that are free from questioning as there were in the 1950s and 1960s. Tropical geography is faced with the problems of modernisation whereas the regional model, which is desired in this field, tends to steer towards the traditional world: so tropical geography is exposed to critics – geographers of the Third World who dream of developing more appropriate models. It is thus the whole field which is in need of restructuring, bearing in mind the lessons of anthropology, sociology or political sciences (Bonnemaison, 1979; Doumenge, 1980) and the detailed ecological and economic studies which are becoming increasingly fruitful (Allaire, Phipps and Stoupy, 1973; de Félice-Katz, 1980). The particular problems of tropical studies recede faced with the urgent need to develop a better articulated and more general view of the subject.

In physical geography the pursuit of more and more specialised objectives has been demonstrated: this is logical but it has become increasingly difficult to define the difference between the contributions of quaternary geologists and morphologists, between the work of meteorologists and specialists of dynamic climatology, between the work of oceanographers and marine geographers. This is not the place to criticise this development, but it seems reasonable to question whether these disciplines should belong to geography. In the short term their development is impeded because the money which they can obtain by forming part of the section of the social sciences is not in accordance with their needs; in the longer term a number of physicists have been asking if it is not time to rethink their discipline and to develop those aspects of it which are useful for the interpretation of human phenomena. Many people are thus interested in the ecological dimension of problems. Work has been started, following the original directions suggested by G. Rougerie (1969), G. Bertrand and O. Dollfus (1973), and J. Tricart and J. Kilian (1979), but not enough of this type of work has been carried out. Physical geography has not been split, although there is growing argument concerning the areas which need to be singled out if more coherence is to be given to the whole.

Anglo-Saxon Attitudes and French Resistance

The modernising of geographical research has largely occurred through the imitation and adaptation of themes initially expounded elsewhere. Inspiration has been sought in Anglo-Saxon work, but also in the other social sciences: the openings in this respect are much more substantial than they have been over the last 20 years.

As far as methodology is concerned, at least a fifth of geographers, especially the young ones, have acquainted themselves with the techniques and methods of quantitative analysis (Ciceri, Marchand and Rimbert, 1977; Beguin, 1979). The procedures are almost the same as elsewhere, with some minor differences; systematic use, as a method of factorial ecology, of Benzecri's correlations (1973), for example, and recourse by the spatial economists of the Claude Ponsard school and the Belgian geographers from Louvain-la-Neuve to graph theory and to the theory of fuzzy sets (Huriot, 1977; Tran, 1977; Beguin and Thisse, 1979).

Amongst the models that Anglo-Saxon research brought into fashion in the 1960s, those which referred to the forces of attraction and to the urban hierarchy have had the most success (Dalmasso, 1971; Pumain, 1982; Saint-Julien, 1982), but more often than not they have simply been used rather than criticised or improved. It is indeed rare to find work such as Bernard Marchand's, which relates directly to the dominant preoccupations of the new Anglo-Saxon geography (1978).

What has caused these differences? A viewpoint that has not yielded to the Anglo-Saxon one? We are grateful to them for having modernised the analysis of feelings of attachment and for having generalised this by talking about perception problems. We are obliged to them for having demonstrated the seriousness of social problems in large modern towns – they do affect French agglomerations. Is the concern to find adequate political answers that will ensure spatial equity foreign to French geographers? No (Reynaud, 1981).

What is maintaining a distance between the French school and the other national schools when most of the obstacles separating them have been removed? The idea that geography can only be articulated around a central theme and that humanist trends, even if they are useful and necessary, do not offer a complete solution. It is necessary to understand spatial linkages. French geography has not renounced its ambition to become a general geography in order to form a better understanding of what structures landscapes and nations (Figure 2.1c). That explains why most geographers envisage themselves studying the arrangement of space (Pinchemel, 1977). We must not forget that this scheme does not have the same positivist or economic connotations that it has in America or Britain. It inspired Juillard's work (1974) and gave value to Roger Brunet's research in the field of objective structures (1972, 1980) and to Armand Frémont's work in the portrayal of lived space (1976).

It is going to take a long time to reconstruct geography around a core of concepts that are relative to spatial organisation (on this theme: Noin, 1976; Dauphiné, 1979; Auriac, 1982). Max. Sorre had already recognised the ecological basis of the way in which space is structured, but his work needs to be updated in view of the fact that he did not have access to modern ecology's great organisational categories – the analysis of energy and matter flows.

In the economic sphere, the rapid progress of geographical knowledge in the 1950s and 1960s was based on two major ideas: firstly, that of the division of labour between economic agents and regions endowed with natural resources, and the skills and equipment which they already possessed; secondly, that of the range of goods which restricts trade to areas of diverse sizes but which are always finite. Problems of specialisation, of the links between complex wholes, and of polarisation, can be explained in this way (Claval, 1963, 1968b). But our societies are not only motivated by the search for economic efficiency – this was even truer of previous societies. Men also fight (or have fought) for this to ensure prestige and power. In the 1970s a large amount of French human geography was thus orientated towards social and political questions.

The relationship between individuals and groups has a spatial expression because it implies the exchange of information (news, orders, supervision and control): these have a price and are subject to limit ranges, whilst channels which maintain these transfers often have a limited capacity. There is thus a geometry of social and political structures, a social architecture (Claval, 1973) which needs to be explored because it plays a large role in the explanation of social organisation. Reductionist notions should not be used in the context of a society and it is necessary to distinguish clearly between objective associations and groups which are conscious of their unity – the only ones who deserve the name of class or region, according to whether the group has a territorial basis or not. Most attention must be focused on systems of social relations which define the dimensions of society. Viewed in this way, political geography ceases to be interested in the state alone (Raffestin, 1980): it describes the power struggle and social regulation which it authorises through simple institutions (family, caste, clientele, pure power, association or exhange) or complex relations (bureaucracy and relations of legitimised power), and the general control which is always exercised by the community (Claval, 1978).

The information used by social communication is of such varying significance that the cost of its transfer is not always constant – it is minimal when there is confidence between the parties concerned. That is what gives more and more weight to the analysis of values and ideologies. The functional study of spatial organisation is linked to the problems of perception, accepted choice, inherited ideals and lived space.

French human geography has always been sensitive to cultural diversity: is cultural diversity not implicit in a description of ways of life? In the 1950s and 1960s, when interest was primarily focused on contemporary problems, it would have been easy to believe that this aspect would lose its attraction; were we not already aware that cultural diversity was fading following the growing Americanisation of the world? A reaction to this superficial point of view has been clearly manifested for the last ten years or so by the experts on the tropics, by specialists on the Middle East and the Far East, and by those people who have remained faithful to historical orientations and to the study of former patterns of territorial organisation.

The fact that it is becoming easier for techniques to spread means that cultures are losing some of their uniqueness. But what is more fundamental than knowing how to do something is the way one lives, thinks and feels. There is nothing to indicate that a levelling is occurring. Geography is thus faced with a new problem: why do men shun the uniformity which is at least within their reach? Why do they fight in order to declare their originality when there is no longer much to differentiate their speech or their equipment? The answer is that they cling to their identity and that spatial organisation has an effect on the way people regard themselves and the way that society regards them (Berque, 1978; Claval, 1981a; Bonnemaison, 1981). Augustin Berque has illustrated this admirably by deciphering the values which led the Japanese to support and to seek high densities (1982).

Thus the new French geography has achieved some measure of success in rethinking itself around the theme of spatial organisation (Figure 2.1c). It has been able to do this by deepening our awareness of the concept of a range, by stretching this to include material goods and communication, and by questioning the meaning implicit in what men exchange as news and what they share as values. Nothing shows this more clearly than the synthetic theories proposed, on tourism some years ago now (Miossec, 1977), on the town more recently (Claval, 1981b).

Systematic studies are dynamic too in other fields: population geography (Noin, 1979), medical geography (Picheral, 1982), services (Labasse, 1974, 1980; Beaujeu-Garnier and Delobez, 1977) and trans-

portation geography (Vigarié, 1979). Cartography is very popular, with
achievements like those of Bertin (1967), and many colleagues are
preparing maps or atlases. Series of regional descriptions, such as the
one directed by Louis Papy, are successful. Many colleagues, specially
in provincial cities, are involved in economic and physical planning.
Some of them are active on the local political scene.

Summary

Writing nearly 20 years ago 'that there is a sense of uneasiness about
present geography', I shocked most of my French colleagues who were
convinced that the discipline was very healthy. Today it is fashionable
to lament the plight of geography and the confusion surrounding the
direction in which it is heading. I think that that appraisal is still an
incorrect one. The crisis is over. French geography is achieving its
original unity and significance, thanks to what it has borrowed from
foreign schools and to its faith in certain original viewpoints. A more
coherent model is replacing the disparate discipline of the 1970s, a
model capable of simultaneously integrating the heritage both of traditional approaches and of the last 20 years. It also offers fresh perspectives on relationships in space, on the individual, on groups, ideologies
and systems of belief: this new geography will be able to interest the
other social sciences and will be in a better position to help men to
shape their environment.

References

Allaire, G., Phipps, M. and Stoupy, M. (1973) 'Analyse écologique des
    structures de l'utilisation du sol', *L'Espace Géographique, 3*, 185-7.
Auriac, F. (1982) *Système économique et espace: un exemple langue-
    docien.* Economica, Paris.
Bailly, A. and Beguin, H. (1982) *Introduction à la géographie humaine*,
    Masson, Paris.
Beaujeu-Garnier, J. (1971) *La Géographie: méthodes et perspectives*,
    Masson, Paris.
———— and Delobez, A. (1977) *Géographie du commerce*, Masson, Paris.
Beguin, H. (1970) *Méthodes d'analyse-géographique quantitative*,
    Litec, Paris.
——— and Thisse, J.E. (1979) 'An axiomatic approach to geographical

space', *Geographical Analysis, 11*, 325-41.

Benzecri, J.P. (1973) *L'Analyse des données*, (2 vol.) Dunod, Paris.

Berdoulay, V. (1981) *La Formation de l'École française de géographie*, Bibliothèque Nationale, Paris.

Berque, A. (1978) 'La Notion de *fûdo'*, *Mondes Asiatiques, 16*, 289-309.

——(1982) *Vivre l'espace au Japon*, PUF, Paris.

Bertin, J. (1967) *Sémiologie graphique*, Mouton, Gauthier-Villars, Paris.

Bertrand, G. (1968) 'Paysage et géographie physique globale', *Revue Géographique des Pyrénées et du Sud-Ouest, 39*, 249-72.

—— and Dollfus, O. (1973) 'L'Himalaya central. Essai d'analyse écologique', *L'Espace Géographique, 2*, 224-32.

Birot, P. (1955) *Les Méthodes de la géomorphologie*, PUF, Paris.

—— (1959) *Précis de géographie physique générale*, A. Colin, Paris.

—— (1965) *Formations végétales du globe*, SEDES, Paris.

Bloch M. (1931) *Les Caractères originaux de l'histoire rurale française*, Institut pour l'Etude comparée des Civilisations, Oslo.

Bonnemaison, Joël (1979) 'Les Voyages et l'enracinement', *L'Espace Géographique, 8*, 303-18.

——(1981) 'Voyage autour du territoire', *L'Espace Géographique, 10*, 249-62.

Boudeville, J.-R. (1961) *Les Espaces économiques*, PUF, Paris.

Braudel, F. (1949) *La Méditerranée et le monde méditerranéen à l'époque de Philippe II*, A. Colin, Paris.

Broc, N. (1975) *La Géographie des Philosophes: géographes et voyageurs français au XVIII° siècle*, Ophrys, Paris.

Brunet, R. (1965) *Les Campagnes toulousaines*, Boisseau, Toulouse.

—— (1968) *Les Phénomènes de discontinuité en géographie*, CNRS, Paris.

——(1972) 'Pour une théorie de la géographie régionale' in *La Pensée géographique française contemporaine*, Presses Universitaires de Bretagne, Saint-Brieuc, 640-62.

—— (1980) 'La Composition des modèles dans l'analyse spatiale', *L'Espace géographique, 9*, 253-65.

Castells, M. (1972) *La Question urbaine*, Maspéro, Paris.

Chevalier, M. (1956) *Les Pyrénées ariégeoises*, Marie-Thérèse Genin, Paris.

Cholley, A. (1942) *Guide de l'étudiant en géographie*, A. Colin, Paris.

Chombart de Lauwe, P.H. (1952) *Paris et l'agglomération parisienne: l'espace social dans une grande ville*, PUF, Paris.

Ciceri, M. Marchand, B. and Rimbert, S. (1977) *Introduction à l'analyse de l'espace*, Masson, Paris.

Claval, P. (1963) *Géographie générale des marchés*, Les Belles Lettres, Paris.

—— (1964) *Essai sur l'évolution de la géographie humaine*, Les Belles Lettres, Paris.

—— (1967) 'Géographie et profondeur sociale', *Annales, économies, sociétés, civilisations, 22*, 1005-46.

——- (1968a) 'La Théorie des villes', *Revue géographique de l'Est, 8*, 3-56.

—— (1968b) *Régions, nations, grands espaces*, Marie-Thérèse Génin, Paris.

——(1973) *Principes de géographie sociale*, Litec, Paris.
253-92.

——- (1976) 'Contemporary human geography in France' in C. Board *et al* (eds.) *Progress in Geography*, vol. 7, Edward Arnold, London, 235-92.

—— (1977) 'Le Marxisme et l'espace', *L'Espace géographique, 6*, 145-64.

——(1978) *Espace et pouvoir*, PUF, Paris.

—— (1981a) 'Les Géographes et les réalités culturelles, *L'Espace géographique, 10*, 242-8.

——(1981b) *La Logique des villes*, Litec, Paris.

Dalmasso, E. (1971) *Milan, capitale économique de l'Italie*, Ophrys, Paris.

Dardel, E. (1952) *L'Homme et la terre; nature de la réalité géographique*, PUF, Paris.

Dauphiné, A. (1979) *Espace, région et système*, Economica, Paris.

Deffontaines, P. (1932) *Les Hommes et leurs travaux dans les pays de la Moyenne Garonne*, Éditions des Facultés Catholiques, Lille.

Derruau, M. (1956) *Précis de géomorphologie*, Masson, Paris.

Dion, R. (1934) *Essai sur la formation du paysage rural français*, Arrault, Tours.

—— (1959) *Histoire de la vigne et du vin en France*, by the author, Paris.

Dollfus, O. (1971) *L'Analyse géographique*, PUF, Paris.

Doumenge, J.P. (1980) *Les Mélanésiens et leur espace en Nouvelle Calédonie*, CEGET, Bordeaux.

Dresch, J. (1941) *Recherches sur l'évolution du relief dans le Massif Central du Grand Atlas*, Arrault, Tours.

Duby, G. and Lardreau, G. (1980) *Dialogues*, Flammarion, Paris.

Dugrand, R. (1963) *Villes et campagnes en Bas-Languedoc*, PUF, Paris.

Félice-Katz, Josée de (1980) 'Analyse éco-énergétique d'un élevage nomade (Touareg) au Niger, dans la région de l'Azawah, *Annales de Géographie, 89*, 57-73.

Flatrès, P. (1957) *Géographie rurale de quatre contrées celtiques Irlande, Galles, Cornwall et Man*, Plihon, Rennes.

Frémont, A. (1976) *La Région, espace vécu*, A. Colin, Paris.

Gallais, J. (1967) *Le Delta intérieur du Niger*, IFAN, Paris.

George, P. (1945) *Géographie sociale du monde*, PUF, Paris.

—— (1951) *Introduction à l'étude géographique de la population du monde*, PUF, Paris.

——et al. (1964) *La Géographie active*, PUF, Paris.

Gourou, P. (1947) *Les Pays tropicaux*, PUF, Paris.

Guilcher, A. (1965) *Précis d'hydrologie marine et continentale*, Masson, Paris.

Huriot, J.M. (1977) *La Formation du paysage économique*, Sirey, Paris.

Isnard, H. (1978) *L'Espace géographique*, PUF, Paris.

—— , Racine, J.B. and Reymond, H. (1981) *Problématiques de la géographie*, PUF, Paris.

Juillard, E. (1953) *La Vie rurale dans la plaine de Basse-Alsace*, Les Belles Lettres, Paris.

—— (1962) La Région, essai de définition, *Annales de géographie, 71*, 483-99.

——(1974) *La Région*, Ophrys, Paris.

Labasse, J. (1955) *Les Capitaux et la région*, A. Colin, Paris.

——(1966) *L'Organisation de l'espace*, Hermann, Paris.

——(1974) *L'Espace financier*, A. Colin, Paris.

——(1980) *L'Hôpital et la ville*, Hermann, Paris.

Lacoste, Y. (1965) *Géographie du sous-développement*, PUF, Paris.

—— (1976) *La Géographie, ça sert d'abord à faire la guerre*, Maspéro, Paris.

Lebeau, R. (1955) *La Vie rurale dans les montagnes du Jura méridional*, Institut d'Études Rhodaniennes, Lyon.

Lefebvre, H. (1974) *La Production d'espace*, Anthropos, Paris.

Le Lannou, M. (1949) *La Géographie humaine*, Flammarion, Paris.

Lipietz, A. (1977) *Le Capital et son espace*, Maspéro, Paris.

Lojkine, J. (1977) *Le Marxisme, l'état et la question urbaine*, PUF, Paris.

Marchand, B. (1978) 'A dialectical approach in geography', *Geographical Analysis, 10*, 105-19.

Meynier, A. (1958) *Les Paysages agraires*, A. Colin, Paris.

Miossec, J.M. (1977) 'Un modèle de l'espace touristique', *L'Espace géographique*, 6, 41-8.

Monbeig, P. (1952) *Planteurs et pionniers de Sao Paulo*, A. Colin, Paris.

Noin, D. (1976) *L'Espace français*, A. Colin, Paris.

—— (1979) *Géographie de la population*, Masson, Paris.

Pédelaborde, P. (1957) *Le Climat du bassin Parisien*, Marie-Thérèse Génin, Paris.

Pélissier, P. (1966) *Les Paysans du Sénégal*, Fabrègue, Saint-Yrieix.

Perroux, F. (1950) 'Les Espaces économiques', *Économie appliquée*, 3, 225-44.

Phlipponeau, M. (1960) *Géographie et action. Introduction à la géographie appliquée*, A. Colin, Paris.

Picheral, H. (1982) 'Géographie médicale, géographie des maladies, géographie de la santé', *L'Espace géographique*, 9, 161-74.

Pinchemel, P. (1977) 'Géographie, espace et organisation de l'espace', *Geographia Polonica*, 36, 173-7.

Planhol, X. de (1958) *Le Monde islamique. Essai de geographie religieuse*, PUF, Paris.

—— (1968) *Les Fondements géographiques de l'histoire de l'Islam*, Flammarion, Paris.

Ponsard, C. (1955) *Économie et espace*, SEDES, Paris.

Pumain, D. (1982) *La Dynamique des villes*, Economica, Paris.

Raffestin, C. (1980) *Pour une géographie du pouvoir*, Litec, Paris.

Reynaud, A. (1974) *La Géographie entre le mythe et la science*, Travaux de l'Institut de Géographie de Reims, Reims.

—— (1981) *Société, espace et justice: inégalités régionales et justice socio-spatiale*, PUF, Paris.

Rochefort, M. (1960) *L'Organisation urbaine en Alsace*, Les Belles Lettres, Paris.

Rougerie, G. (1960) *Le Façonnement actuel des modelès en Côte-d'Ivoire forestière*, IFAN, Dakar.

—— (1969) *Géographie des paysages*, PUF, Paris.

—— (1975) *Les Cadres de vie*, PUF, Paris.

Roupnel, G. (1932) *Histoire de la campagne française*, Grasset, Paris.

Saint-Julien, T. (1982) *Croissance industrielle et système urbain*, Economica, Paris.

Sautter, G. (1966) *De l'Atlantique au fleuve Congo, une géographie du sous-peuplement*, Mouton, La Haye, Paris.

Sorre, M. (1943-52) *Les Fondements de la géographie humaine* (4 vol.), Paris, A. Colin.

—— (1957) *Rencontres de la géographie et de la sociologie*, Marcel Rivière, Paris.

Tran, Qui Phuoc (1978) *Les Régions économiques floues. Application au cas de la France*, Sirey, Paris.

Tricart, J. and Cailleux, A. (1953-65) *Traité de géomorphologie*, 1°part, *Géomorphologie climatique*, vol. I-V.

—— and Kilian, J. (1979) *L'Éco-géographie*, Maspéro, Paris.

Vigarié, A. (1979) *Ports et vie littorale*, Hachette, Paris.

Zeldin, T. (1973-7) *France 1848-1945* (2 vol.), Clarendon, Oxford.

3 ITALY

Berardo Cori

Strong Quantitative Progress

Between the late 1940s and the first years of the present decade remarkable quantitative progress was witnessed in Italian geography, in almost every aspect of it and in parallel with the increase in the university student population.

The number of university institutes totally or partially concerned with geography or its various branches increased from around 25 to 60. The number of chairs in geographical disciplines rose from less than 20, in three faculties alone, to around 75, distributed among the Faculties of Arts, Education, Economics, Science, Architecture, Political Science and Languages. Courses of a geographical nature offered in these faculties increased from 75 to 350. 'Geographers', as physical persons, of whom there were 88 in Italy according to the *World Directory of Geographers* published in 1952, numbered almost 500 on the basis of an assessment made by the Association of Italian Geographers around 1980. Membership of the Florence Society of Geographical Studies increased from a mere 87 in 1946 to the 1981 figure of 552.

There was also an increase in the number of journals and periodicals of a geographical character appearing alongside the basic *Bollettino della Società Geografica Italiana* and the *Rivista Geografica Italiana*, and also in the number of publishers who 'discovered' geography as a publishing sector with an expanding market. Consequently, there was a marked increase in the number of publications by Italian geographers and on geographical subjects: consulting the two comprehensive bibliographies for this century (*Società Geografica Italiana*, 1964; *Associazione dei Geografi Italiani*, 1980) we are able to establish a number of around 500 for the pre-Fascist period (1901-22), over 1,300 for the Fascist period itself (1923-43), almost 2,000 for the immediate post-war period (1944-61), and more than 3,200 for the last 20 years (1962-80).

One fact alone is in contrast with the others which, together, suggest an 'irresistible upward trend': participation of Italian geographers at international geographical congresses. At the Lisbon Congress (1949)

16 Italian geographers were present and read 13 papers, whilst in Tokyo (1980) there were 19 Italians with 17 papers. Practically no development occurred in this area, therefore, and I don't believe that the greater distance of Tokyo compared with Lisbon was a determining factor.

Evolution and Stability in Research Interests

It is reasonable to ask whether, in such a vast and varied discipline as geography, quantitative development has been accompanied by evolution in the preferences of scholars for the various branches of the subject.

The answer differs according to whether we consider the large traditional divisions of geography or the smaller systematic divisions based on individual areas of investigation. In the case of the first, two facts appear to be evident: the substantial stability of physical geography; and the differing fortunes of economic geography on the one hand and of historical geography on the other. Around one-quarter of all geographical studies carried out in Italy have concerned physical geography, both before and after the Second World War; if anything at all has changed, it has been the disciplinary background of the authors, among whom we find, to an increasing extent in recent times, geologists rather than geographers. Economic geography, on the other hand, has witnessed enormous growth in popularity this century — in Italy as well as elsewhere — whereas there has been a declining interest in historical geography *lato sensu* (i.e. including the history of geography, the history of cartography, etc.): in the pre-Fascist period fewer than one in ten geographical studies dealt with economic matters, whilst more than four in ten had an historical bias; today, studies in economic geography represent one-quarter of the total and those in historical geography less than one-sixth. A downward trend occurred in the historico-geographical disciplines, therefore, and an upward one in economic geography, even though its peak was reached in the 1950s rather than during the last 20 years. The main body of Italian geographers would seem to be composed of a minority with a stable attraction towards physical geography and a majority generally orientated towards human geography, but whose interests have radically changed over the course of the century, away from the 'historic' towards the 'contemporary'.

If we turn our attention to the 'thin sections' within the discipline, we need to mention geomorphology, urban geography and industrial geography among those sectors showing rapid development since the Second World War. These three branches, together with population geography, represent those which have received most attention during the post-war decades in Italy. After a long period of stagnation, geomorphology began to develop again, especially after 1960, due mainly to the growing interest shown by geologists in the subject, and it has therefore quite naturally been taking on the character of a highly specialised and autonomous discipline. The increase in urban and industrial geography represents a logical and positive response, if still often superficial, to the processes of industrialisation and urbanisation which have characterised Italy since the war. Hand in hand with this increase has been the fall in interest in geographical studies on agriculture and rural settlement, which had shown signs of revival between 1945 and 1960, as a result of phenomena of intellectual inertia and of national economic policy at the time, based extensively upon agriculture and exemplified in the substantial attempts at agricultural reform in southern Italy. Far more clear-cut and, one would suggest, practically definitive has been the decline in studies on political geography, which in the 1930s and early 1940s had grown to great importance under the influence of Fascist colonial and expansionist policy and of German geopolitics. And it is precisely to this twofold negative image that such studies owe their almost complete disappearance. The history of cartography and the history of exploration (this latter for reasons partly similar to those affecting political geography) have been those sectors contributing most to the relative decline of the historico-geographical branches of our discipline.

The Last Classics

When geographical research activity resumed as early as 1944-5 after the wartime interlude, the scene was dominated by a small group of men, born in the last quarter of the nineteenth century, who had taken Italian geography to a level approaching that reached in the most advanced European nations.

This generation had been raised in the school of some of the great nineteenth-century scholars, such as Giuseppe Dalla Vedova (1834-1919) and Giovanni Marinelli (1846-1900) and in the wake of the heroic era of African exploration. The group had prematurely lost its principal exponent, the son of Giovanni Marinelli, Olinto (1874-1926), and its work was hindered towards the end of the 1930s by the racial discrimination which deprived another of its leaders, the Jewish scholar Roberto Almagià (1884-1962), of his university chair. All the same, the value of these men — most inspired by the positivist school of German origin, partially modified by Vidalian possibilist influences, but with others continuing in an old national historico-erudite tradition — was responsible for the achievement of substantial results in the period before the Second World War.

Olinto Marinelli and above all Giotto Dainelli (1878-1968) had been protagonists in some of the last ventures of scientific exploration of the earth: in East Africa (Eritrea and Danakil, 1905-6; the Lake Tana region, 1936-9), and to a greater extent in Karakorum, in Tibet and in the Himalayas (1913-14 and 1929-30). In 1922 Marinelli produced his *Atlante dei tipi geografici*, the first systematic work of a comparative and integral reading of Italian geographical landscapes through the interpretation of topographical maps, later to be imitated more than once outside Italy. In 1940 Dainelli produced his *Atlante fisico-econ- omico d'Italia*, one of the first examples in the world of national atlases. In the field of history of cartography, a fundamental work, *Monumenta Italiae Cartographica*, was published by Almagià in 1929, whilst in the sector of history of exploration a number of important works appeared: of particular importance were those on Columbus and Vespucci by Giuseppe Caraci (1893-1971). In physical geography the name of Antonio R. Toniolo (1881-1955: *Compendio di geografia generale*, 1940) stands out, and in human geography the figure of Renato Biasutti (1878-1965), geographer, anthropologist, ethnologist, with his work *Le razze e i popoli della Terra* (1941) – a work completely alien to racial prejudice, despite the period in which it was written. Umberto Toschi (1897-1966) introduced studies in urban geography and in theoretical economic geography to Italy in the 1930s; these had been little practised till this period.

Several of the above geographers had acquired positions of interna- tional prestige: Almagià was awarded the Malte-Brun prize by the Paris Geographical Society in 1923; Dainelli won the international competi- tion organised by Egypt for appointment to the chair of geography at Cairo in 1929; Biasutti was made honorary member of the German Geographical Society and of the British and German anthropological societies.

It was the 1930s which marked the climax of this good season of Italian geography, notwithstanding the increasing weight of condition- ing from the regime and the unfortunate, passive conformity to Fascist directives of too many Italian geographers: the 'years of consent' had arrived. During this period, with the affirmation of promising young geographers, such as Elio Migliorini (born 1902) and Aldo Sestini (born 1904), seemingly giving hope for future developments in Italian geography, and with the new university degree course in geography, introduced in 1924, in full swing, a conspicuous attempt was made, especially on the part of Toniolo, Biasutti, Almagià and others (amongst whom was Ferdinando Milone, born in 1896), to create a new organisation of geographical research based on the collective, systematic and also interdisciplinary study of a series of major themes. The passage from spontaneous and individual research to organic, group research — fundamental in any modern discipline — was therefore attempted (even if in the hierarchical fashion of the times) by the geographers of this period, with financial support from the newly founded National Research Council, one of whose organs was the Committee for Geography.

Results followed quickly: between 1933 and 1940 publication commenced on several series of studies on variation of beaches, on marine and fluvial terraces, on the altimetric distribution of vegetation, on historic variation in climate, on rural dwellings (preceded by Biasutti's well-known paper given at the international congress in Paris in 1931), on the depopulation of mountain regions, and on Italian cities and ports. As can be seen, these were central themes in Italian geography, a number of which touched upon crucial problems for the nation's development (even though such themes were those welcomed or at least tolerated by Fascism and though the solutions proposed by the regime were never questioned); the themes were studied using a modern, uniform and co-ordinated methodology which proved to be suitable for the production of homogeneous results in the 30 volumes printed before the outbreak of war.

After the upheaval caused by the war, therefore, Italian geography appeared in a lively and active state and relatively advanced in terms of the integralist-possibilist-idiographic paradigm which by then was almost universally accepted. It was through the very esteem which Almagià enjoyed internationally that he was elected vice-president of the International Geographical Union (IGU) at the first post-war congress in 1949, despite the fact that at this time Italy was still kept on the fringe of international life. Right up until the end of the 1950s Italian geography continued to live off the work of its masters. And these, even after retiring from active teaching (Biasutti and Dainelli in 1948, Almagià in 1954), continued to produce works of importance. Dainelli, with age and infirmity, was no longer able to travel, but wrote a number of extremely valid works of synthesis, such as *La conquista della Terra* (1950). And it was to important works of synthesis that others dedicated themselves: Biasutti, to his famous work *Il paesaggio terrestre* published in 1947; and, although considerably younger, Toschi, to his *Geografia economica* (1959), and Milone, to his *L'Italia nell'economia delle sue regioni* (1955). Caraci, more closely involved in research in a strict sense, published his *Italiani e Catalani nella primitiva cartografia nautica medievale* (1959). Almagià continued to be dominant in the field of studies on the history of cartography (*Monumenta Cartographica Vaticana*, 1944-52), but also turned his hand to synthesis and popularisation in the field of general and regional geography with *Fondamenti di geografia generale* (1945-6), the powerful volumes of *Il mondo attuale* (1953-5) and *L'Italia* (1959). In 1952 he was awarded the American Geographical Society's 'Cullum Geographical Medal' and his medal cabinet was completed in 1958 with the 'Victoria Medal' of the Royal Geographical Society.

The 1950s also saw the resumption, and in several cases the completion, of those series begun in the 1930s and interrupted by the war. The National Research Council, in which Almagià authoritatively defended the interests of geography, created four new research centres: for human geography (directed by Almagià himself), for physical geography (Toniolo), for ethnological geography (Biasutti) and for economic geography (Colamonico), all four of which were to assume responsibility for continuing the series of publications and for setting up others. In conclusion, everything seemed to indicate that the Italian school of geography had now firmly established itself among those of international repute.

A Generation which Fell out of Step

As often happens, the years in which success seems to become established are also the years in which stagnation sets in, in which development ceases. It is true that the great 'end-of-the-nineteenth-century' generation was still alive and active (with the exception of Toniolo) around 1960. It is also true that some members of this generation continued to teach (Caraci until 1953, Milone until 1966) and to produce works of distinction: Dainelli's *Le Alpi* in 1963, Toschi's *La città* in 1966 and Caraci's admirable refutation of the authenticity of the 'Vinland Map' in 1967 (which his American colleagues were to arrive at only several years later). It is further true that some of those born at the very beginning of the century can be considered as continuers of the previous great generation: and here I allude to Sestini in particular, whose *Il paesaggio* appeared in 1963 as the last valid product of traditional geography. And, finally, it is true that certain series of systematic studies were continued and brought to completion even after 1960 (the series on rural dwellings, for example, with its total of 29 volumes) and that other series were begun: in 1956, the series on land use in Italian regions (although at the time of writing it has still not been completed) and in 1960 the series of regional monographs (also fostered by Almagià, and furthermore popular rather than scientific in nature).

At the same time, however, several clearly negative facts emerged. Some, the least serious among them, were organisational: restructuring of the National Research Council, which led to the suppression in the late 1950s of three of the four geographical research centres (the fourth surviving for another decade), and in 1963 the disappearance of the old Committee for Geography absorbed into a plethoric committee dealing with all the humanities. Other, more serious, facts were those connected with the generational change-over. The 'great generation' continued, as I have said, to produce things of value, but not necessarily to produce and to teach *new* things; this was a natural effect of its ageing. *Neither*, unfortunately, did the new generation, born roughly between 1905 and 1925, whose members began to occupy university chairs immediately after the war, and during the 1950s moved in alongside the old generation to control and direct Italian geography, finally replacing it in the 1960s (Almagià died in 1962, Biasutti in 1965, Toschi prematurely in 1966, and Dainelli, nonagenarian, in 1968).

Let's make things quite clear, however: we are not talking about an incompetent generation. In physical geography, for example, an honourable tradition was continued and developed in geomorphological and climatological studies, especially in the Universities of Padua and Pisa. Ortolani's field-work in Asia and South America and the theoretical work of Bonetti, Gribaudi and Massi stand as examples of the valid continuation of other noble traditions in Italian geography. Stimulating new ideas came from an outsider, Francesco Compagna (1921-82), a journalist and politician as well as a geographer. It should be noted, however, that men like Compagna were looked upon with diffidence and kept in the wings for many years by the geographers 'who mattered'.

The crux of the matter was this: with several exceptions, the incoming generation was conformist and rather unoriginal. Although no longer having an alibi provided by the regime, it acted as if one still existed and as if geography had ceased to evolve with the basic acquisitions of Vidalian thought transmitted and imposed by the previous generation. The fervour of ideas of the past was replaced by a lifeless repetitiveness which continued to reproduce the old positivist or erudite formulae *ad infinitum*, and confused an organic and systematic approach to research with avoidance of any innovatory stimulus, and even of a surrogate form of innovation such as keeping in touch with international geographical literature (which in the meantime was evolving rapidly and in a stimulating manner). The idiographic paradigm was entering into crisis, and this generation (1) not only did not realise this fact, but (2) applied idiographic rules in a particularly dull way (differently from what happened in other countries which consciously remained faithful to the said paradigm) and (3) did its best to impose it as an article of faith.

It is paradoxical, moreover, that alongside the conformism of this generation there existed another, equally harmful characteristic which the masters of the previous generation had attempted to avoid in the organisation of research work during the 1930s, and that was isolationism. This was pointed out by Almagià shortly before his death when he decried — albeit in an article praising the state of Italian geography — 'a certain spirit of isolationism and individualism which still inspires many of our scholars and which induces them to prefer personal inquiry to work in collaboration' (1961, p. 431).

How this situation came about is not easy to explain. It was certainly determined to some extent by the reorganisation of geography in secondary education, following the Gentile reform act of 1924, which limited both recruitment channels and professional outlets for geographers (Capel, 1977). Responsibility must also be attributed to the climate of cultural autarchy under Fascism and the rigid hierarchical tutelage that the scholars of the previous generation exercised over those of the following one (Almagià could be called, as it has been said for Vidal in France, an 'intellectual autocrat'). Whatever the reasons were, with conformism and isolationism the generation of Italian geographers produced in the first quarter of the twentieth century followed a directly opposite course to that which Giotto Dainelli had recommended to geographers, 'unlimited curiosity, great spirit of order . . .'.

The Case of Lucio Gambi

In the general atmosphere described above, there appeared between 1956 and 1964 a series of pamphlets, later brought together in an anthology (Gambi, 1964) and subsequently followed by others, also united in a single text (Gambi, 1973), the work of one of the youngest geographers of the generation I have just accused of having 'fallen out of step'. Lucio Gambi (born 1921) stood out in this generation through his anticonformism, his sound humanistic culture, his capacity for thinking, producing new ideas and writing with style, and his ability to engage vigorously in controversy: qualities which in all truthfulness should not be so rare among university professors and yet in the discomforting panorama of his contemporaries stood out as exceptional. As a result, whether because they finally represented something new and original, whether because they emerged as valid in the dullness of the cultural situation in which they were launched, or because of the sensation they caused, Gambi's ideas deserve to be briefly summarised here in that they are central to the history of post-war Italian geography.

In the first place Gambi stigmatised the obtuseness of the majority of Italian professors of geography at the time, depicting them as culturally backward, out of harmony with most modern currents of thought, lacking in critical energy and innovatory impulse. The world of Italian geography 'gives one the impression of an old fallow field, or even better, of a cultural area cut off from the great currents in which culture circulates, is nurtured and evolves' (1964, p. 123). It has an enclosed world where 'every current of fresh air is feared' (1964, p. 88), where 'there is extreme and moreover uncritical respect for the ways of interpreting science that the older generation had formulated', where 'sterile canons of mere descriptive notionalism' were accepted, and where work was carried out according to 'patterns and paradigms of childlike crudeness' (1964, p. 94). In brief, Italian geography around 1960 presented itself substantially – as the title of one of the pamphlets pointed out – as a 'depressed area'. Read again *today* such claims appear perhaps somewhat excessive, but not unfounded, and are anyway shared by a good number of geographers of the succeeding generation. Made *at that time*, however, with Italian geography indulging in rites of self-glorification, they were to many truly upsetting, like the child's observation of the truth in the tale of *The Emperor's New Clothes*.

This serious cultural inadequacy depended to a great extent, according to Gambi, on the old unitary conception of geography matured and developed in a positivist climate which Gambi questioned, decried and ably destroyed with sound argument. His ideas on this theme had elements of both obviousness and notable originality. The former can be summed up in the claim that no one could work well, and obtain scientifically valid results, in both physical and human geography, given the diversity and moreover the opposition between methods in the science of nature and the science of man. The latter consisted of a clear proposal for the identification of three separate disciplines in traditional geography:

(a) the analysis and interpretation of the Earth's natural phenomena, or physical geography; (b) the examination of the relations between the environment and organic life on Earth, or ecology; (c) the history of economic conquest and of man's instrumental organisation of the Earth, or human geography. (Gambi, 1964, p. 72.)

The clarity and freshness of this vision (for Italy, at least) and especially of the historically-orientated conception of human geography which it contained, set against the cultural backcloth described earlier, was welcomed with relief and gratitude by the younger geographers, at least inasmuch as it provided a breaking-point with the past and a basis on which to open a new epistemological debate on geography. Its limitations were seen in an insufficient consideration of geography in other countries. For some time Hartshorne had worked on the problem of physical geography in a less radical yet more convincing way, distinguishing its specialist branches, now an integral part of natural sciences, from the study of the environment for man, an indispensable premise and complement to human geography. And at the same time or shortly after, whilst in countries where geographical studies were considered advanced physical geography – in the Hartshornean sense – and human geography continued to be closely related with respect to research, there was a rise in world geography of neopositivism, quantitativism, modelistics, systems analysis, perception geography, behaviourism, all potentially 'unifying' forces within geography, bringing with them, if nothing else, new forms of dualism (Dematteis, 1970, pp. 56-61).

In the light of the stimulating challenges of these new geographies, much of the successive development in Gambi's thinking seems irrelevant or frankly disappointing. His participation in the student movement of 1968 (Gambi, 1968) was ingenuous and uncritical; his choice of position, classist and Marxist (Gambi, 1973, pp. ix, 71), was clear, although it seemed not to have any necessary epistemological implications (cf. sect. 8); his anti-quantitativism presented as anti-imperialism (1973, p. 76) was unacceptable. More interesting, on the other hand, and more likely to lead to considerable developments if adequately followed up, are his thoughts on 'disciplines' and 'problems':

> Reality is not divided or broken up into disciplines, which are con-
> structs founded on conventional concepts ... ; reality is made up of
> facts, and the facts become the subject of science when they create
> problems ... Disciplines don't exist as elements of cultural structure
> of the society, but only as instruments of such. And for society,
> there exist problems to be solved with whatever means of knowledge
> available, within the organic nature of their terms. (Gambi, 1973,
> p. 75.)

Among the problems that society faces today, Gambi recognised a vast
collection which can be generically defined as 'organisation of space',
studied by geographers as by a variety of other scholars with different
disciplinary labels. And with this unitary view, the tripartite division of
geography with the radical oppositions in methods originally advanced
by Gambi begins to lose some of its value.

Much Ink about Nothing

In the last 20 years, as we saw in the first section of this chapter, the
written production of Italian geographers has reached a quantitatively
high level. This abundance, however, appears disproportionate in
respect of the qualitative value of the same. The generation which 'fell
out of step' in fact, apart from still being active and from having given
little thought to the themes of Gambi's criticism — either peevishly
rejected or condescendingly played down — is silently engaged in the
large-scale activity of self-reproduction tending to perpetuate, through
the normal mechanisms of competitive examinations and co-opting,* a
traditional conception of geography which has nothing at all to do with
respect for the disciplinary tradition. Such respect is always positive,
but the best way to show it is by helping geography to progress.

Self-reproduction, on the other hand, tends to preserve those
harmful aspects of individualism which Almagià has decried.

The organisation of geographical research in Italy is still traditional,

---

*Even the author of this note (born 1938) has no hesitation in recognising that he
is a product of this 'Operation Xerox' and therefore has his part to play in the
defects which are stigmatised here.

that is spontaneous and individual and conducted using artisanal techniques . . . The situation most commonly found is that of the professor surrounded by his pupils who assist him in his research. Interdisciplinary work is usually the result of completely fortuitous encounters. (Valussi, 1973, p. 34.)

As a consequence of this atmosphere, the work of bringing the old series of geographical studies to completion is a difficult one, and the very few and ephemeral new ones organised are in the hands of study commissions which often exist only on paper or consist of the arithmetical sum of a number of individuals.

From the simple nonexistence of real, modern team-work, it is not uncommon to find behaviour of a contentious and conflictual character, which in one respect gives rise to 'not infrequent feuds between individuals or "schools" and *coups de main* in attempts to assert individual power' (Corna-Pellegrini, 1980b, p. 906) and in another respect, in a more serious and general way, prevents or limits the formation of organic departments of geography in the larger Italian universities, as envisaged by the 1980 University Reform Act (and thus geography is deprived of the possibility of obtaining adequate structures and funding for research). It also obstructs the realisation of a modern national atlas (and so Italy, one of the precursors in this field in 1940, thanks to Dainelli, is now in the 1980s the only developed nation without a national atlas).

In Italian geography during the 1960s and 1970s little was studied and little read (especially of international literature). On the other hand a great deal was written (by individuals: collaboration is still looked upon with suspicion and counts little in competitive examinations!) Scientific discussion at congresses was weak and impoverished in content, whilst formalist and organisational diatribes were interminable, with an intolerable expenditure of energy and intellect. Theoretical, methodological and epistemological debate was more intelligent, and yet even this at times appeared excessive — as Almagià (1961, p. 421) had emphasised — to those who believe, along with Hartshorne, that 'we shall not learn geography by discussing how geography itself should be studied'. This was all the more true as there emerged a notable disproportion between the amount of discussion and agreement on the one hand and of application on the other; those approaches which were most original and had a greater cultural base were more often quoted than put into practice. Gambi was on the lips and in the footnotes of a good number of geographers, yet rarely is his influence seen in the content of their work.

If there was no lack of writing on minor themes, there was a scarcity of both important works of synthesis and work dedicated to substantial problems which Italy faced. Baldacci's treatise on general geography (1972) says nothing new and moreover represents a retrograde step in respect of Almagià's manual (1945-6); and even the description of Italy edited by Pinna and Ruocco (1980) cannot compare (being composed of contributions from a variety of authors of mixed value and showing no evidence of co-ordination) with Almagià's (1959). And what of the major national problems? Whilst the 'demand' for territorial knowledge increased strongly, through the progressive entanglement of a series of knots in the ecological, energy, industrial, demographic, regional and town-planning sectors, 'national problems today appear to be felt very little or not at all by geographers' (Gambi, 1973, p. 33) or were tackled with approaches which were inadequate for the complexity and importance of the problems themselves. There were, of course, many exceptions: but on the whole work that should be typically that of the geographer was being conducted to an increasing extent by economists, town planners, sociologists, etc., acting as substitutes, or else by *foreign* geographers (one need only think of the problems of the two main Italian metropolises tackled with distinction by two French geographers – see Dalmasso, 1978 – and hardly touched upon by Italians).

'Anyone well up with geographical studies in other countries cannot remain indifferent before the vivid spectacle of our inferiority . . . , of our disconcerting provincialism.' It was in these terms that one of the most authoritative exponents of the generation born in the first years of the century, Gribaudi, addressed a meeting of geographers in 1965. The well-known Tricart Report, published shortly afterwards, fully confirmed this assessment of Italian geography. Yet awareness of inferiority is evidently not sufficient to induce the older geographers to update their own knowledge or that of their students.

The most striking example of this gap, of the cultural backwardness of Italian geographers, is provided by the fortunes of the quantitative-theoretical revolution. This revolution, as is known, asserted itself internationally in the early 1960s, but was only 'discovered' in Italy in 1970, and even then by a young geographer (Dematteis, 1970). Far from arousing that critical interest which should naturally awake at the arrival of new ideas, this discovery seems mainly to have annoyed the majority of Italian geographers who for some considerable time rejected, *a priori* and uncritically, the new quantitative paradigm, without serious thought or reasoned confutation. Examples of serious critical analysis of new geography and reasoned opposition to it *have* appeared in Italian journals – but at the hands of foreign geographers (George, 1971; Claval, 1976). When some scholars eventually began to get to grips with new geography (and not all through serious conviction; rather because it was fashionable) they did so belatedly, at a time when in other countries it was in its rethinking phase. And the anti-quantitativism or a-quantitativism, still dominant in Italy, does not participate in this rethinking *a posteriori* (Turco, 1980b, p. 876) but, more simply and more sadly, in an atmosphere of general ignorance and immobility.

The consequences for geography teaching of such a situation – student indifference, inadequate preparation of secondary school teachers, the anachronism of the degree course in geography and contempt from the rest of the faculty – are easily imagined, as is the powerful feedback effect of such consequences.

Geography and Planning: an Appointment Missed

In these conditions of peevish isolationism, cultural backwardness and narrow-minded insensitivity towards the problems of society – conditions not without exceptions but which characterise the majority, especially of the geographers who matter – it is no surprise that applied or active geography in Italy has few followers. Consequently, 'in our geographical journals there is no section dealing with those national and regional laws, the application of which often calls for considerable effort in acquiring territorial knowledge' (Dematteis, 1980, p. 486) and 'the community of geographers carries little weight in the vast movement which is developing in the field of control and management of territory, both at national and regional level' (Turco, 1980b, p. 870). To little avail was the lucid preliminary study by Nice, *Geografia e planificazione territoriale*, written in 1953 and appearing well before the classic works of Labasse, Phlipponeau, etc. To little avail was the presence and keen stimulus of a front-line politician, such as Compagna, among geographers. Looked upon with diffidence (as is all that is new or thought to be new) by the older generation (with some exceptions: Toschi, Merlini), never to become a part – since never taught – of the mental framework of the younger generation, contested at times by exponents of geographical radicalism as enslavement to power, participation of Italian geographers in *solving* the problems of the country or its regions, on whatever scale, has been even less than their participation in *studying* such problems.

Very few geographers are found in an active role on the territorial, urban, economic, etc. programming and planning committees which are proliferating in Italy. And the few that do take part do so because, by chance, they are known to have some specific competence or because they are active in one of the political parties, in which case they are part too of the widespread party political subdividing of every kind of public office. Substitution of the work of the geographer, and the assumption of the functions of territorial science, comes extensively — given the almost total lack of response from qualified geographers — not only from disciplines which do effectively have a territorial orientation, but also from sciences such as sociology, economics and others with much less of a territorial mentality.

Not even this is sufficient to rouse our average geographer. His reaction usually oscillates between whimpering victimisation, half-hidden satisfaction at the frequent cases of failure in programming and behaving like the fox towards the unreachable grapes in the classical fable.

Marxist Geography or Marxist Geographers?

The reaction to nothingness often brings about an exasperated quest for something new. And it is in this context that I believe we can evaluate the birth in the 1970s of a Marxist trend in Italian geography. There is evidently a parallel — given the difference in proportions — with the rise of radical geography as a response to certain aspects of the crisis in international geography. Italian Marxist geography, however, has little connection with the contemporary radicalism of Harvey or Peet, little with Gambi's historicist criticism before it, and even less with developments in Soviet geography.

It is related more closely with the anarchism of Lacoste and the *Hérodote* group, drawing inspiration from the 'transposition into the field of geography of those ideas and concepts which a few years earlier neo-Marxist economists, planners and sociologists had begun to apply to the multi-disciplinary analysis of territory' and also from 're-reading Marx in search of useful principles and categories of research for modern geography' (Dematteis, 1980, p. 783). This latter course, intellectually quite fruitful, characterises above all the work of Massimo Quaini (born 1941) and in particular his book about Marxism and geography (1974). Here, Quaini (pp. 2-4), after denouncing the 'lack of logic and scientific rigour which seem not to be eliminable from the geography passed on by the previous generation', after defining new geography as 'a revision which in many cases affects the language used rather than content or scientific method', and after denouncing 'the relationship it has with the decision-making centres of endorsement and subordination to choices they make', asserts that 'Marx's criticism . . . already contains essential aspects of that "new" and more rigorous logical instrumentation needed in order to establish a science of territory which is neither chaotic erudition nor a kind of apologia'.

Quaini's reflections gave birth to development and liaison among geographers who see themselves as exponents of this line, developments such as the journal *Hérodote/Italia* with its declared Lacostian inspiration (1978; in 1982 it changed its name to *Erodoto*) and the association, 'Democratic Geography', which held its first pithy congress in 1979 (Geografia Democratica, 1981). With certain of these developments epistemological considerations are to some extent exceeded by politico-praxist ones. 'To see the clash between old and new only in terms of scientific paradigms can be deforming . . . The contraposition of old and new cannot but have clear political implications . . . The discussions formulated in this way become necessarily classist' (Quaini, 1978, pp. 152-3.) For this reason the *Hérodote/Italia* manifesto (Quaini, 1978, pp. 158-9, 162) emphasises the need 'to think space in order to organise struggle', proclaims itself 'against the geography of the Pentagon', and declares its tendency towards a 'communist reappropriation of territory'. Even as early as 1975, geographers of Marxist inspiration had tabled a motion at the Salerno Geographical Congress which was exclusively political and totally devoid of any epistemological connotations, echoing uncritically Communist theses on terrorist violence in Italy and praising the 'success of the people of Indochina in the bitter struggle for freedom against American imperialist capitalism' ('Cronaca del Congresso', 1977, pp. 69-70). It would be interesting to see the evolution in the thinking of the signatories of that motion after the Pol-Pot massacres in Cambodia, the tragedy of the boat-people and the emergence of Vietnamese imperialism; yet it is even more interesting to note that, among the signatures on the motion, whilst Quaini's is missing, we find those of some of the most old-fashioned exponents of geographico-traditionalist nothingness . . .

In the face of the excess of a 'current of thought which claims for geography the totally new role of denouncing the misdeeds of the capitalist system, according to an interpretation which frequently considers the ideological base more important than factual evidence' (Corna-Pellegrini, 1980b, p. 906), if the behaviour of certain traditionalist geographers is the expected neo-conformism and *'gattopardism'*, the behaviour of an innovatory geographer, Compagna, takes the form of a condemnation of the 'fumes of ideological intoxication which in this happy land has spared no one' (Compagna, 1980, p. 898). And this said, the resumption of serious dialogue is still possible (Vallega, 1979a; Turco, 1981).

Meanwhile, however, we might ask, as does Celant (1980, pp.720-3), if we can in fact speak of a new 'Marxist geography' in Italy or simply of a number of geographers professing Marxist ideology. Celant's answer leaves no doubt:

> The attempt to connote Marxist ideology scientifically represents an aim for the scholars involved . . . but judging by the amount of Italian geographical material which has been published so far, it constitutes an objective which must be considered to have been postponed . . . In geography, with the present state of scientific elaboration, Marxism is an ideological model in the sense that it offers a foundation for political action; but it is not a paradigm . . . One can practise Marxist ideology yet at the same time adhere to a quantitative or possibilist paradigm; ideological practise and adherence to paradigms are . . . facts independent one from the other.

Da Pozzo (1979, p. 106) speaks even more bluntly:

> If rather than a new essence of science, substitution of the class that makes use of it is being proposed, then . . . it is more correct to recognise that revolutionary and conservative geography don't exist, but rather geographers belonging to one or other of the sides.

The other leading Marxist geographer, Giuseppe Dematteis (born 1935), likewise has to agree with this and it is with his words (1980, p. 788) that we can for the moment bring the discussion to a close: 'I don't believe that we can, at least for the moment, in Italy, speak of a tendential Marxist paradigm in geography . . . '

Timid Attempts to Get Back into Step

Despite all that has been said so far, it is perhaps possible to conclude on a cautious note of optimism. We shall not get back into step with international geography by trying at all costs to do something new, by chasing frenzied dreams of cultural nationalism, by practising the art of sprinting into the lead or that of superficially respraying old tools and hurdling over or belittling whole phases or themes in the scientific evolution of our discipline. Gambians and Marxists have not created new paradigms, but they have had the merit of unmasking the fossilisation of the old ones, a state which has been perpetrated for a long time and even presented as a means of 'defending geography', whilst all the world was advancing under the banner of tumultuous and fruitful 'progress in geography'.

We shall only get back into step by catching up with the international geography train as it steams along, by coupling on to the most vital and fruitful of its many-coloured carriages and by guiding our own research on to its advanced track. I believe that if we put our minds to it we can succeed, since in the last few years something in the panorama described so far has changed.

There are new forces: even if they still lack a group of renewers like the French *équipe* of *L'Espace géographique*, the unyielding barrier of competitive examination self-reproduction has already been broken through by a number of talented and original thinkers, born in the early 1930s, reaching university chairs. Another unit of younger men, born in the early 1940s, yet equally valid, is advancing as reinforcement.

There are now, to a greater extent than in the past, means of attenuating what has euphemistically been called 'the gap in cultural interests between the national and international scientific communities' (Turco, 1981, p. 130): today, Italian geographers travel more (and a little less *en touriste*), read more books and journals from abroad, take advantage, through certain journals, of the scientific consultancy of authoritative foreign geographers and have a considerable number of works in translation available, works by French, British, American and German geographers which, now found extensively in student reading lists, compensate for the limitations and inadequacies of our own original production.

Gambian demystification, the new forces, the revivified contribution of the best exponents of previous generations and new instruments are already having a certain positive effect under our very eyes. Some works of synthesis which appeared in the late 1970s (Pinna, 1977; Castiglioni, 1979) have resumed the tradition of 'major works' which seemed to disappear with Toschi, and have shown at the same time that physical geography can still be done and done with excellence, by *geographers* rather than by geologists, meteorologists or other specialists. Other important works are in an advanced stage of preparation, from a universal geography, extensively revised in content, to a national thematic atlas which will finally — thanks more to the Italian Touring Club than to geographers — fill the well-noted gap. Greater involvement of geographers is being seen in the study of problems of planning, organisation of territory, the *Mezzogiorno*, the relationship between city and industry, regions, ecology, protection of cultural and environmental wealth, as well as in solid participation in initiatives promoted by state bodies, local authorities and sectorial organisations. Interdisciplinary dialogue — see the case of the 1982 round table between geographers and town planners on the settlement system in Central-Southern Italy — seems to have become less reserved, more open and free of inferiority complexes, especially now that there is greater awareness of the fact that other social disciplines suffer from inadequacies and disfunctions not too dissimilar to our own. Theoretical work is becoming more frequent and, what matters most, more pithy (see especially the *Rivista Geografica Italiana* over the years from 1979 onwards), whilst the epistemological debate between the various tendencies, if still a little long-winded, reached a high level at the Varese Congress of 1980 (Associazione dei Geografi Italiani, 1980) and at the Trieste meeting of 1982 (Pagnini, 1982). Finally and above all, there is a breakthrough into Italy, with reasoned application and discussion, of the most advanced geographical themes and methodologies, from quantitative geography (Vagaggini and Dematteis, 1976; Pagnini and Turco, 1982) to perception geography (Geipel and Cesa-Bianchi, 1980; Corna-Pellegrini, (ed.), 1980) and systemic geography (Vallega, 1979b, 1982).

Along these guide-lines, in my opinion, we must continue with constancy. And we must continue as well to develop a new and more correct scientific-professional code of ethics, which is maturing even if it is certainly not yet part of the behaviour of the majority of Italian geographers, and which consists in repudiating the narrow-mindedness, the dogmatism, the methodological rigidity, the fossilisation, the lack of self-criticism and renewal, the anathemas ('this isn't geography!'), the cultural pecking-order based on age and academic status and, in the last analysis, repudiating all that prevented the generation of Almagiàs and Biasuttis (and their guilt is here not light) from producing another equally valid one, and that has allowed this second generation to hinder — partly with success — the formation of a better third one.

References

Almagià, R. (1961) 'La geografia in Italia dal 1860 al 1960', *L 'Universo*, *41*, 419-32.

Associazione dei Geografi Italiani (1980) *La ricerca geografica in Italia 1960-1980*, Ask, Varese.

Baldacci, O. (1972) *Geografia generale*, Unione Tipogr. Edit. Torinese, Torino.

Biasutti, R. (1934) 'Ricerche sui tipi degli insediamenti rurali in Italia', *Comptes Rendus Congr. Int. Géogr., Paris 1931*, A. Colin, Paris, *3*, 7-16.

Bonetti, E. (1975) *Principi di geografia economica generale*, Cluet, Trieste.

Capel, H. (1977) 'Institucionalizacion de la geografia y estrategias de la comunidad cientifica de los geografos. El caso de la geografia italiana', *Geo-Critica, 9*, 1-26.

Caraci, G. (1967) 'Il falso del secolo: la Vinland Map', *Bollettino della Società Geografica Italiana, 104*, 178-214.

Castiglioni, G.B. (1979) *Geomorfologia*, Unione Tipogr. Edit. Torinese, Torino.

Celant, A. (1980) 'I paradigmi nella ricerca geografica' in Associazione dei Geografi Italiani, *La ricerca geografica in Italia 1960-1980*, Ask, Varese, 713-28.

—— (1982) 'Entropia e concentrazione della popolazione', *Rivista Geografica Italiana, 89*, 3-53.

Claval, P. (1976) 'La Brêve Histoire de la nouvelle géographie', *Rivista Geofrafica Italiana, 83*, 395-424.

Compagna, F. (1967) *La politica della città*, Laterza, Bari.

—— (1980) 'Figlioli miei, geografi immaginari' in Associazione dei Geografi Italiani, *La ricerca geografica in Italia 1960-80*, Ask, Varese, 897-902.

Cori, B. (1976) 'Avviamento a un dibattito sulle strutture e lo svolgimento dei congressi geografici italiani', *Rivista Geografica Italiana, 83*, 257-86.

Corna-Pellegrini, G. (ed.) (1980a) 'Numero monografico dedicato a "Geografia e percezione dell'ambiente" ', *Rivista Geografica Italiana, 87*, 1-132.

—— (1980b) 'La geografia italiana e la società moderna' in Associazione dei Geografi Italiani, *La ricerca geografica in Italia 1960-1980*, Ask, Varese, 903-8.

'Cronaca del Congresso' (1977) in *Atti XXII Congr. Geogr. Ital., Salerno*

1975, Ist. Graf. Ital., Cercola, *1*, 43-99.

Dalmasso, E. (1978) 'La Recherche géographique française en Italie depuis la fin de la deuxième guerre mondiale', *Mél. Ecole Franc. Rome, 90*, 15-33.

Da Pozzo, C. (1979) 'Recensione a Quaini, M. (1978)', *Rivista Geografica Italiana, 86*, 105-8.

Dematteis, G. (1970) *'Rivoluzione quantitativa' e nuova geografia*, Laborat. Geogr. Econ., Torino.

—— (1980) 'La risposta dei geografi ai problemi di conoscenza posti dallo sviluppo della società italiana' and 'La nascita dell'indirizzo marxista nella ricerca geografica italiana' in Associazione dei Geografi Italiani, *La Ricerca geografica in Italia 1960-1980*, Ask, Varese, 483-90, 781-92.

Gambi, L. (1964) *Questioni di geografia*, Ediz. Scient. Ital., Napoli.

—— (1968) *Geografia e contestazione 1968*, Lega, Faenza.

—— (1973) *Una geografia per la storia*, Einaudi, Torino.

Geipel, R. and Cesa-Bianchi, M. (eds.) (1980) *Ricerca geografica e percezione dell'ambiente*, Unicopli, Milano.

Geografia Democratica (1981) *L'inchiesta sul terreno in geografia*, Giappichelli, Torino.

George, P. (1971) 'La Géographie quantitative, un nouveau déterminisme?', *Notiz. Geogr. Econ., 2*, 33-43.

Gribaudi, D. (1963) 'Contro una critica demolitrice della geografia', *Rivista Geografica Italiana, 70*, 245-70.

Lacoste, Y. and Coppola, P. (eds.) (1977) *Crisi della geografia, geografia della crisi*, F. Angeli, Milano.

Massi, E. (1972) 'Analisi regionale e sviluppo polarizzato', *Bollettino della Società Geografica Italiana, 109*, suppl., 71-108.

Merlini, G. (1969) 'La geografia applicata: esperienze e prospettive' in *Atti XX Congr. Geogr. Ital., Roma 1967*, Soc. Geogr. Ital., Roma, *2*, 417-598.

Pagnini, M.P. (ed.) (1982) *Teoria e misura dello spazio geografico*, Trieste.

—— and Turco, A. (1982) 'Quantitative and Theoretical Geography in Italy' in P.J. Bennett (ed.), *European Progress in Spatial Analysis*, Pion, London.

Pinna, M. (1977) *Climatologia*, Unione Tipogr. Edit. Torinese, Torino.

—— and Ruocco, D. (eds.) (1980) *Italy, a Geographical Survey*, Pacini, Pisa.

Pracchi, R. (1976) 'Avviamento a un dibattito sulla geografia generale', *Rivista Geografica Italiana, 83*, 1-28.

Quaini, M. (1974) *Marxismo e geografia*, La Nuova Italia, Firenze.
——(1978) *Dopo la geografia*, Espresso-Strumenti, Milano.
Sestini, A. (1962) 'La geografia nell'insegnamento universitario' in *Atti XVIII Congr. Geogr. Ital., Trieste 1961*, Ist. Geogr. Univ., Trieste, *1*, 483-507.
Società Geografica Italiana (1964) 'Un sessantennio di ricerca geografica italiana', *Mem. Soc. Geogr. Ital., 26*, 1-632.
Turco, A. (1980a) 'L'Emploi des modèles dans l'analyse des problèmes territoriaux en Italie' in *Les modèles comme source d'inspiration dans la géographie contemporaine*, Inst. Géogr. Univ., Lausanne, 43-73.
——(1980b) 'I modelli nei paradigmi della geografia italiana' in Associazione dei Geografi Italiani, *La ricerca geografica in Italia 1960-1980*, Ask, Varese, 865-80.
——(1981) 'Classici della geografia, quantitativismo e possibilità di riunificazione dei paradigmi disciplinari', *Rivista Geografica Italiana, 86*, 129-52.
Vagaggini, V. and Dematteis, G. (1976) *I metodi analitici della geografia*, La Nuova Italia, Firenze.
Vallega, A. (1979a) 'Neopositivismo e marxismo in geografia: riflessioni su un dibattito', *Rivista Geografica Italiana, 86*, 129-52.
—— (1979b) 'Porti e regionalizzazione: un paradigma sistemico', *Bollettino della Società Geografica Italiana, 116*, 577-94.
—— (ed.) (1982) 'Numero monografico dedicato alla "Regionalizzazione" ', *Rivista Geografica Italiana, 89*.
Valussi, G. (1973) 'Ricerca geografica e insegnamento universitario' in *Atti XXI Congr. Geogr. Ital., Verbania 1971*, Ist. Geogr. De Agostini, Novara, *2*, 4, 7-240.

4 SOUTH-EAST EUROPE

Gyorgi Enyedi and Adam Kertesz

This chapter deals with geography in five countries — Bulgaria, Czechoslovakia, Hungary, Romania and Yugoslavia — in all of which it has a long history. Universities were established in the Czech and Hungarian kingdoms several centuries ago, but the present socio-political structure of the area was only established in 1920, until when the whole of Czechoslovakia and much of Yugoslavia were part of the Austro-Hungarian Empire. In the two previously independent states — Hungary and Romania — geographical societies were established in 1871 and 1875 respectively (Bulla, 1954; Mihailescu, 1976), and independent geographical institutes were founded in the 1870s at the universities in Budapest, Prague, Brünn, Pressburg and Zagreb.

The development of geography at this time arose from roots in geology and in statistics (defined then as the description of states). Close relations with geology have been maintained to the present day. Physical geography is prominent throughout the area, especially geomorphology, and geography is usually regarded as one of the natural sciences. In teacher training institutions, too, it is more common to connect geography with the natural sciences (especially biology) than with history.

Before 1945 the content of geography in Czechoslovakia and Hungary was strongly influenced by the prevailing German school, whereas elsewhere the major influence was from France. The classical German school was particularly influential on the development of geomorphology, whereas in human geography the *anthropogeographie* and *géographie humaine* traditions flourished, to the detriment of studies in economic geography. There was little influence from either the Soviet Union or the Anglo-Saxon countries.

Since 1945 geography has enjoyed a considerable expansion, with chairs established in many universities, a growth in student numbers and the foundation of new journals. Growth was most rapid in the less-developed Balkan countries, and the number of geography departments in Hungary in fact declined, by one. In the 1940s and 1950s, too, independent geographical institutes were established under the auspices of national academies of sciences (Pecsi, 1976; Mazur, 1978); following the Soviet lead, these academies were reorganised into networks of research institutes operated by the state. Their major function (relative to that of a university) was to foster research, and many of the leading scientists left the universities, thereby separating themselves from the teaching role. (The geographical institute of the Romanian Academy of Sciences was closed in the 1970s, however, and united with the geographical institute at the University of Bucharest.)

During this period since 1945 the expansion of geography has been accompanied by an enhancement of the reputation of geographical research and of the role of geography in public life. This greater prestige is related to the close bonds developed between geography and regional planning, regional economics and, later, the study of natural resources and the planning of environmental control. Planning benefits from the synthetic geographical viewpoint, and so in several countries, such as Hungary, geographical research institutes have been active in directing primary research aimed at promoting regional development and natural resource utilisation. At the same time, however, the role of geography in the public education system has been reduced. It has not been dissolved but the number of classes in the subject has decreased, particularly at secondary schools, and geography is no longer taught at certain types of school. This has had a particularly adverse effect in countries where the main role of the universities is to train teachers (although some who have graduated as trained teachers now work in regional planning institutions). Training of applied geographers has been undertaken at only a few of the universities.

Until the late 1950s geography in these countries (except Yugoslavia) was overwhelmingly influenced by Soviet work, much of which was made available in translation for the first time. Economic geography was very largely theoretical and involved introducing a Marxist theory of space; there was little development of research methodology. Soviet physical geography was much more advanced methodologically, however, and was not as biased towards geomorphology.

During the first 15 years after the Second World War there was a sharp separation of physical and economic geography accompanied by internal specialisation within each. The specialised researchers sought relationships with other disciplines in the natural and social sciences rather than with each other, so that the geographical research institutes abandoned their previous human/landscape concepts and became multidisciplinary workshops, cohesive only by convention and by a common focus on space. This break with the past was so complete that it was not until the 1960s that the merits of pre-war geography were realised and the earlier foci were integrated with the more modern work. In part, this break reflects the small size of the pre-war generation of geographers, which was decimated both during the war and by emigration immediately after. The 'new' geography after 1945 was represented by a new, young generation, with all the enthusiasm and 'one-sidedness' of iconoclasts.

The break with the synthesising focus contradicted the trend in all of the countries to focus attention on the minutely-detailed description of available natural and social resources. Voluminous geographical monographs were compiled in Bulgaria and Romania – *Geografija na Balgarija* (Balgarija Akademija, 1966, 1961) and *Geografia fizica a Republicii Populare Rominel* (Calinescu *et al.*, 1955) – and the whole of the area was described at a meso-regional level. Because the concept of landscape had been abandoned, there were separate treatments of physical and economic geography, in encyclopedic fashion, with chapters on the various branches but no geographical synthesis.

Trends in Physical Geography

Geomorphology

As already indicated, geomorphological research has dominated the physical geography literature in these countries. Initially, regional geomorphological studies were undertaken, of, for example, the evolution and landforms of a valley or mountain range (Lang, 1955; Adam, Marosi and Szilard, 1959). These began with palaeogeomorphological surveys, to establish the geochronology, and were followed by descriptions of the present landforms. The latter included descriptions of the natural and anthropogenic processes shaping the contemporary landforms (mass movements, soil erosion, wind deflation, etc.). Later works supplement such *qualitative* treatment with *quantitative* elements. Morphometric features were analysed (such as relative relief, hypsometric curve, shape conditions), to aid the morphographic description of regions. More recently still, laboratory analyses of materials – including radio-carbon dating and palaeomagnetic investigations – have supplemented the regional descriptions.

Classical geomorphological investigations remain significant (Georgiev, 1976; Sifrer, 1978; Mazurova, 1978). But the regional descriptions of the 1950s are being extended by more specialist work aimed at revealing the complex geomorphological evolution. The crucial role of streams has thus been recognised, with studies of alluvial terraces, floodplains and alluvial plains (Milic, 1977).

During the 1960s geomorphologists developed interests in other topics, such as tropical peneplanation and pedimentation (i.e. the genetic investigation of planated surfaces). One specialism which developed out of classical geomorphology, but with close links to geology, was *Quaternary research*. It began with the study of periglacial phenomena and processes (Czudek and Demek, 1961; Demek, 1977; Pecsi, 1963; Zeremski, 1977; Ichim, 1980) and of Quaternary sediments. Genetic and stratigraphical investigations of loess and loess-like sediments now use up-to-date methods, such as palaeomagnetic investigations, thermal analysis and radiocarbon dating (Bognar, 1978; Pecsi, 1982). In addition to stratigraphical-chronological topics, research on loess includes analyses of its chemical, physical and lithological properties, denudation and the relationships between geomorphological processes and loess formation.

In general, research on geomorphological processes is now more concerned with the mechanisms themselves than with observations and descriptions of their consequences. Thus topics relating to stream erosion and deposition, slope processes and erosion by wind have come to the fore, and in the last decade their understanding has been advanced through an increased amount of field and laboratory experimental work (Zajicek, 1981).

Applied Geomorphology

Much of the theoretical research discussed above was stimulated by social demands, which have also generated an increased volume of directly practical, applied geomorphology (Morariu, 1980; Demek, 1980). This latter-trend reflects the important requirement of applicability which is attached to research in socialist countries.

Initially, the applied work — begun in Czechoslovakia and Hungary in the early 1960s — involved geomorphological mapping, representing in graphical form the results of research into geohistory. The regional morphological pattern, together with the nature of the underlying rocks and the age of the landforms, were the elements commonly depicted. The detailed mapping enriched geomorphological research, with its requirements for information that had to be represented. A synthesis was a crucial part of this work, involving both qualitative and quantitative representation of landforms.

Practical demands for regional planning, engineering geology and construction geology have required engineering geomorphological mapping as well as genetic geomorphological mapping. These new methods were developed by Soviet geomorphologists, and have mainly been adopted in Hungary (Pecsi, 1971; Demek, 1972). And to serve the' interests of engineers research has been extended into the study of mass movement on slopes and of landslides with work on classifying landslides and their trigger mechanisms (Urbanek, 1968) and on mapping the distribution of landslide hazards (Pecsi and Juhasz, 1974).

Contemporaneously with the growth of interest in geomorphological mapping came research into anthropogenic processes (Radalescu, 1972). The various man-induced processes in different countries and regions were investigated. Particular attention was paid to agricultural lands; the processes of soil erosion were identified, and patterns of actual and potential erosion were mapped (Stehlik, 1981).

In summary, applied geomorphology has strengthened in recent years, alongside work in the classical tradition. The latter is character-ised by increasing specialisation on the one hand but by the multi-faceted approach to problems on the other; a somewhat contradictory situation in which specialisation is complemented by a holistic view. The development of applied geomorphology has been promoted by its disçlosure of anthropogenic processes, phenomena and landforms through traditional methods, and advanced by its contribution to the solution of practical problems in the field of environmental control.

Other Fields within Physical Geography

Hydrogeographical research has mainly been involved in the mono-graph descriptions of regions (e.g. Daneva, 1975; Ujvari, 1980), con-tributing palaeohydrological investigations of drainage networks and catchment areas to complement palaeogeomorphological studies (Somogyi, 1961). The focus was on water discharge, erosion by surface water and water supply; sub-surface waters, groundwater, mineral and thermal waters have mainly been treated by hydrologists (except in Czechoslovakia, where several geographers have been involved in such work). The instrumentation of small catchments has been a major geo-graphical contribution, providing a hydrological input to studies of geo-morphological and pedological topics, many concerned with soil erosion and its amelioration, particularly with regard to agriculture.

A common topic in both hydrogeography and geomorphology is the study of Karst, which has its richest traditions in Yugoslavia. It covers the speleology, geomorphology, ecology and cenology of Karst areas.

Phytogeographical research has been undertaken to provide data for geobotanical maps that indicate an area's potential vegetation. Other studies have related vegetation to relief and other landscape elements. Studies of plant and soil geography are usually incorporated into regional landscape monographs. In addition, several Czechoslovak workers have tackled problems of pedogenesis and the classification of soils and their zonality. In Hungary, apart from detailed analyses of soils in representative model areas, geographers have evaluated the agricultural potential of sites (Goczan, 1971).

Landscape and Region

The most important goal of all work in physical geography is regionalisation. Geomorphological and landscape regions have been delimited for each of the countries (e.g. Mazur and Luknis, 1978; Saru, 1980), partly as contributions to separate regional studies and partly as components of, or bases for, national atlases.

In recent decades, ecological studies have gained ground, bringing investigations of landscapes and their evaluation to the foreground. The characterisation of landscape types by their ecological contents and economic value provides basic material of the potential of each type. As a result, research is and will be concentrated on the landscape and its evolution.

Economic Geography

Investigation by Economic Sector

In economic geography, the separate study of the various sectors or branches has become dominant. Geographical investigations provide descriptions of the location of particular branches of production, including the manufacture of single products. (Many of the relevant papers emphasise practical planning points, but practising planners have rarely taken much notice of these papers.)

In the 1950s poor statistical data made thorough regional analyses impossible. Thus, although this was the decade of the first wave of industrialisation in several of the countries and their regions, and major changes in spatial structure took place as a consequence of industrial location decisions, researchers were forced to concentrate on population and agrarian geography (Sarfalvi (ed.), 1966; Radalescu, Velcea and Petrescu, 1968; Friganovic, 1970; Stefanescu, 1975; Bernat and Enyedi, 1977). The focus was on production. There was a supposition that progress in consumption (and thus in the tertiary sector) will follow automatically from that in the productive sectors. Thus geographical investigation of non-productive sectors was almost unknown: population was studied only as a source of labour; research into settlement geography was surpassed; and social and cultural geography were virtually non-existent.

Studies in economic geography were intended to demonstrate the superiority of social laws in determining the geographical distribution of population. A break with physical geography, previously the leading element within geography in South-east Europe, was a necessary ideological task. An overemphasis on social laws led to a denial of the influence that the natural environment exerts on production; 'geographical nihilism', as it was labelled locally, spread as a result.

Since the 1960s compartmentalisation into branches has become much less rigid, and the scope of research has widened considerably. Industrial geography remains relatively neglected, however, with few studies of the distribution of industry. Population and agrarian studies remain popular, in the context of the wider social process of industrialisation, which has both stimuli and consequences far beyond industrial location (Kluna, 1955; Ivanicka, 1964; Christov, 1975; Popovici, Crangu and Manesco, 1977; Bora, 1978; Vriser, 1981).

Investigations of settlement geography — the transformation of the settlement pattern and the functions of settlements — have broadened in scope remarkably. Urban growth attracts most attention and rural problems are in the background, although the countries of South-east Europe are to a large extent still rural (Tufescu, 1972; Beluszky, 1973; Kivadshiev, 1977; Christov, 1977; Anderle, 1978; Cucu and Deica, 1978; Vresk, 1978). It is not only the productive forces stimulating rapid urbanisation that are studied, but also the infrastructural and social implications. Concepts and methods for such work have been updated, and a typological trend (the establishment of geographical types) has become prominent (Ivanicka, Zelenska and Mladek, 1966; Enyedi, 1980; Gosar, Nihevo and Jakos, 1980).

The geography of tourism is also becoming a distinct typical study. Historical geography is active only in Romania, however, where it is subject to political goals of doubtful value. Expressly social and cultural geographical publications are sparse, while political geography (unable to break free from recollections of the ill-famed German geopolitics) is virtually non-existent.

Economic Regions

A main development in Soviet economic geography, influential in South-east Europe, has been the so-called school of rayons. Led by its outstanding founder, N.N. Baransky, this school believes that economic life is objectively arrayed into special units characterised by the complexity of productive branches and a certain level of specialisation. The economic regions (*rayons*) differ in their production structure, and it was assumed that the nature of the settlement network and of living conditions follows automatically from changes in production. The complexity of the regions reflected the relative absence of long-distance inter-regional links in the gigantic Soviet space economy, so that each region was basically autarkic.

To apply this autarkic concept to the small territory of South-east Europe was impossible. Consequently, there were sharp debates over the economic regionalisations of the countries, which did not fully support the theory (Markos, 1952; Mihailescu, 1970; Rogic, 1973; Krajko, 1977). Economic regionalisation was basic to the introduction of a planned economy in the 1940s and 1950s, however. This included regional development planning. It was believed that the administrative structure formed in the Middle Ages was inappropriate for the direction of the economy, since the old administrative units obviously did not coincide with the new economic regions. Since, in a direct economic management system, the economy is governed through the administration, it was a general view that economic regionalisation should serve as the basis for a reformed regional administration. This has been carried out in all of the countries (in some more than once), except Hungary. The regions are not autarkic units, however, and the reforms had political rather than economic goals, because the economy is managed by sectoral ministries rather than by local government bodies.

The study of economic regions has declined in its interest as a research topic. The problem of complex (non-sectoral) units has not been solved and, mainly because of the absence of statistical data, research has been confined within the framework of administrative units.

Applied Economic Geography

Much economic geography research has been conducted with some economic planning objective, primarily the optimal regional allocation of production by central economic management bodies. This has brought geographers into contact with economists and planners; regional economics developed as a separate discipline only in the 1960s.

The economics of production could only be studied as an optimising exercise incompletely, because cost-benefit analyses are meaningless given the arbitrary prices fixed by the direct central management system. In this context, however, work by geographers involved the survey, and to a certain degree the evaluation, of natural resources.

This practical orientation means that geographical research is exclusively national, if not local, in character. Research abroad is not undertaken, and the teaching of foreign countries at the universities has no research foundation. Within the various countries, the benefits economic decision-makers gain from geographical research depend more on the mechanisms of decision-making than on the quality of the geographers' work. Their greatest influence appears to be in Hungary and Yugoslavia, although regional planning is generally important in the local decision level of the planning systems.

Since the 1960s and the end of the Cold War geographers in South-east Europe have been more open to overseas influence. Modern research trends initiated in the Anglo-Saxon countries have been particularly influential, though less so than in Poland. Hungarian and Yugoslav geography are perhaps most open to such influences, including those of German and French workers. The impact of changes in Soviet geography has been substantial, for obvious reasons, and knowledge of quantitative methods has frequently been obtained via Soviet mediation. Regarding scientific co-operation within the International Geographical Union (IGU) and in the organisation of international conferences and bilateral meetings, the Czechoslovakian and Hungarian geographers have been most active, with consequences for the national research efforts.

New Research Trends

Quantitative Geography

The quantitative revolution reached South-east Europe from the English-speaking countries and the Soviet Union only after some delay. Some researchers were applying mathematical and statistical methods in the 1950s (Enyedi, 1957) but they had little influence, partly due to the poor mathematical qualifications of geographers but also to the practical orientation of much work which had an aversion to over-abstract models that did not work in practice. New methods were adopted mainly to improve the standard of traditional research accomplishments (in typologies and regionalisation, for example).

Quantitative methods have been used mainly in the study of the regional structure of the economy (Popov, 1972, 1976; Sikos, 1977; Zuljic, 1977) and in geomorphology. In the latter, Horton's and Strahler's methods of network analysis were adopted (Krcho, 1964; Kertesz, 1972), followed by morphometrical methods of map analysis. Theoretical approaches to the study of landscape description and the interrelationships among landscape factors were chiefly taken up by Czechoslovak workers (Krcho, 1974). The application of mathematical-statistical methods in a systems theory context has been restricted mainly to Hungary and Czechoslovakia (especially Slovakia); they have been applied by a few individuals in Yugoslavia and Bulgaria, and appear to be practically unknown (as far as one can tell from publications) in Romania.

Geography as Environmental Systems Analysis

From the 1960s on the separation of physical and economic geography has become less rigid, following the arguments of the Soviet geographer Anuchin for a uniform geography (1960). This gave rise to practical research in the field of landscape evaluation (Marosi and Szilard, 1974), assessing the value of the landscape in terms of socio-economic benefits. The sectoral approach, within which the geography of agriculture was prominent, laid an increasing emphasis on the economic function of natural resources.

In the early 1970s, mainly due to the influence of another Soviet geographer, Gerasimov, the trend was towards integrated environmental research in which society is represented as a single large system. This is divided into natural and social systems, which are in turn subdivided into the physical environment, the transformed environment, the economic sphere and the cultural sphere (and further, finer-grained subdivisions). Adherents of this approach see it as an up-to-date research concept for geography, strengthened by the growing attention given to environmental management and control. It is a reformulation of the traditional concept of a unified geography, set in the context of systems theory. Examples of research in this context, revealing causal processes, are few, however, and the pragmatic work of the last decades in service of society has thrust theoretical work well into the background.

Summary

It is not the geography of South-east Europe that is cultivated in the five countries. The interests of geographers are nationally introverted. Excepting university texts of traditional character, the economic geography has been treated only twice in recent decades (Straszewicz, 1974; Enyedi, 1978), and the physical geography not at all. And within two of the countries there are divisions reflecting the ethnic framework.

In Czechoslovakia, the Czech and Slovak universities and research institutes are independent of each other, whereas in Yugoslavia each member republic has its own academy of sciences, university institutes and separate publications; there is no national Yugoslav geographical journal, therefore, and local publications are issued infrequently and irregularly.

References

Adam, L., Marosi, S. and Szilard, J. (1959) *A Mezofold termeszeti foldrajza* (Physical geography of the Mezofold), Akademiài Kiado, Budapest.

Anderle, A. (1978) 'Vyvoj ceskoslovenskych mest 1868-1970', *Geografisky Casopis, 30*, 2, 126-50.

Anuchin, V.A. (1960) *Teoreticheskie problemi geografii*, Nauka, Moskwa.

Balgarija Akademija (1966, 1961) *Geografija na Balgarija* vol. I. *Fizicseszka geografija*, vol. II *Ikonomicseszkaja geografija*, Szofija.

Beluszky, P. (1973) 'A telepulesszabalyozas nehany elvi-modszertani szempontja', *Foldrajzi Ertesito, 22*, 4, 453-66.

Bernat, T. and Enyedi, Gy. (1977) *A magyar mezogazdasag teruleti problemai* (Regional problems in Hungarian agriculture), Akademiai Kiado, Budapest.

Bognar, A. (1978) 'Les i lesu slicni sedimenti Hrvatske'. *Geografski Glasnik, 40*, 21-40.

Bora, Gy. (1978) 'The stages of development in the industrial system of Budapest' in M.K. Bandman (ed.), *Projection of the formation of territorial production complexes*, USSR Academy of Sciences, Novosibirsk, 24-41.

Bulla, B. (1954) 'Nehany szo a magyar foldrajztudomany halado hagyomanyairol' (On the Progressive Traditions of the Hungarian Geographical science), *Foldrajzi Kozlemenyek, 78*, 1. 1-10.

Calinescu, M. *et al.* (eds.) (1955) *Geografia Fizica a Republicii Populare Rominel*, Litogr. Invatemintului, Bucaresti.

Christov, T. (1975) 'Geografia na promishlennosta v Blgarija v usloviyata na sotsialisticheskata ikonomicheska integracii', *Problemi na geografiata, 1*, 3, 18-29.

—— (1977) 'Rayonno-teritorialna klasifikacia na selishtata v Blgaria', *Problemi na Geografiata, 3*, 1, 59-64.

Cucu, V. and Deica, P. (1978) 'Le processus d'urbanisation en Roumanie', *Revue Roumaine de Géographie, 22*, 2, 239-51.

Czudek, J. and Demek, J. (1961) 'Vyzman pleistocenni krioplanace na vyvoj povrchovych tvaru Ceske vysociny', *Anthropos, 14*, 57-69.

Daneva, N. (1975) 'Kolitchestven analiz na nadlshnite profili na viakoi reki Severoistotchna Blgaria', *Problemi na geografiata, 1*, 2, 17-33.

Demek, J. (1972) *Concept and Methods of Geomorphological Mapping*, Academia, Prague.

—— (1977) 'Periglacialni geomorfologie: soucasne problemy a vychlidky do boudoucnosti', *Zpravy Geografickeho Ustavu CASV, 14*, 4, 73-91.

—— (1980) 'The geographical prognosis in present-day Czech geography', *Sbornik Ceskoslovenske Spolecnosti Zemepisne, 85*, 1, 3-8.

Enyedi, Gy. (1957) *'A mezogazdasagi korzetek kutatasanak uk modszerei'* (New Methods in the Research of Agricultural Regions), Mezogazdasagi Kiado, Budapest.

—— (1978) *Kelet-Kozep-Europa gasdsagi fildrajza* (Economic geography of East-Central-Europe), Kozgazdasagi es Jogi Kiado, Budapest.

—— (1980) 'Regional types of rural living conditions in Hungary' in: Gy. Enyedi and J. Meszaros (eds.), *Development of Settlement Systems*, Akademiai Kiado, Budapest, 205-17

Friganovic, M. (1970) 'Gravitacijske zone dnevne migracije u radne centre Hrvatske', *Geografiski Glasnik, 23*, 89-99.

Georgiev, M. (1976) 'Otnosno geomorfoloshkata interpretacia na geoloshkata osnova v Blgaria s ogled genetishnatz diferenciacia na landshaftite', *Problemi na geografiata, 2*, 1, 13-22.

Goczan, L. (1971) *Soil geography of the Marcal Basin*, Geographical Research Institute, Hungarian Academy of Sciences, Budapest, Abstract no. 17.

Gosar, L., Nihevo, P. and Jakos, A. (1980) 'Pomen tipologije naselij za planiranje', *Geografiski Vestnik, 52*, 63-79.

Ichim, I. (1980) 'Probleme ale cercetarii periglaciarului din Romania', *Studii si Cercetari de Geografie, 27*, 1, 127-36.

Ivanicka, K. (1964) 'Process of industrialization of Slovakia', *Geograficky Casopis, 16*, 4, 215-27.

—— Zelenska, A. and Mladek, J. (1966) 'Funkcionalne typy vidieckych sidiel Slovenska' in K. Ivanicka (ed.), *Aspekty studia regionalnej geografickej struktury*, Slovenske Pedagogicke Nakladelstvo, Bratislava, 51-93.

Kertesz, A. (1972) 'Matematikai-statisztikai modszerek alkalmazasi lehetosegei a geomorfologiaban a Tetves-arok es a Pelivolgy peldajan' (Possibilities of application of the mathematical-statistical methods in geomorphology on the example of the Tetves and the Peli Valley), *Foldrajzi Ertesito, 21*, 4, 487-502.

Kivadshiev, S. (1977) 'Stopanskogeografska klasifikacia na gradovete v Blgaria', *Problemi na Geografiata, 3*, 1, 65-74.

Kluna, A. (1955) *Manufakturni obdobi v Cechach*, Nakladelstvi CSAV, Praha.

Krajko, Gy. (1977) 'A gazdasagi korzetek taxonomiai szerkezet az Alfoldon' (The taxonomic structure of economic regions in the Great Hungarian Plain). *Alfoldi Tanulmanyok, 1*, 80-92.

Krcho, J. (1964) 'Morphometric analysis of the slope conditions of the Kosice Basin', *Acta Geologica et Geografia Universitatis Comenianae, 4*, 1-23.

—— (1974) 'Struktura a piestorova diferenciacia fyzickogeografickej sfery ako kybernetickeho systemu', *Geograficky Casopis, 26*, 2, 133-63.

Lang, S. (1955) *A Matra es a Borzsony termeszeti foldrajza* (Physical geography of the Matra and Borzsony Mts.), Akademiai Kiado, Budapest.

Markos, Gy. (1952) 'Magyarorszag gazdasagi korzetbeosztasa' (Economic regionalisation of Hungary), *Foldrajzi Ertesito, 1*, 3, 582-634.

Marosi, S. and Szilard, J. (1974) 'Landschaftsbewertung und Landschaftsanalyse', *Foldrajzi Ertesito, 23*, 2, 203-6.

Mazur, E. (1978), '25 rokov Geografickeho ustavu Slovenskej akademie vied', *Geograficky Casopis, 30*, 3, 200-18.

—— and Luknis, M. (1978) 'Regionalne geomorfologicky clenenie Slovenskej Socialistickej republiky', *Geograficky Casopis, 30*, 2, 101-25.

Mazurova, V. (1978) 'Terasy riek ceskoslovenskich Karpat a ich uztah k terasam Dunaja', *Geograficky Casopis, 30*, 3, 281-301.

Mihailescu, V. (1970) 'Die Wirtschaftsregionen Rumaniens', *Revue Roumaine de Géographie, 14*, 2, 197-209.

—— (1976) 'L'École géographique roumaine – vue générale', *Revue Roumaine de géologie et de géographie, 20*, 5-15.

Milic, C.S. (1977) 'Osnovne odlike fluvialnog reliefa Srbije', *Zbornik Radova Geografiski Institut, 29*, 1-34.

Morariu, T. (1980) 'Objectifs prioritaires dans la recherche géographique appliquée en Roumanie', *Revue Roumaine de Géographie, 24*, 13-18.

Pecsi, M. (1963) 'Die periglazialen Ercheinungen in. Ungarn', *Petermanns Geographische Mitteilungen, 107*, 3, 161-82.

—— (1971) 'Concept and methods of geomorphological mapping',

American Geographer, 3-4 (1971), 72.

—— (1976) 'Az MTA Foldrajztudomanyi Kutato Intezet 25 eve (1951-1976)', *Foldrajzi Ertesito, 25*, 2-4, 137-74.

—— and Juasz, A. (1974) 'Kataster der Rutschungsgebiete in Ungarn und ihre kartographische Darstellung', *Foldrajzi Ertesito, 23*, 2, 193-202.

—— (1982) 'The most typical loess profiles in Hungary' in M. Pecsi (ed.) *Quaternary Studies in Hungary*, Geographical Research Institute, Budapest, 145-69.

Popov, P. (1972) *Matematicheski metodi i ikonomicheskata geografia*, Izdatelstvo Nauka, Sofia.

—— (1976) 'Sistemi i sistemen analiz v geografiata', *Problemi na geografiata, 2*, 1, 42-50.

Popovici, I., Crangu, A. and Manescu, L. (1977) 'Répartition géographique de l'industrie et développement équilibré des départements de la Republique Socialiste de Roumanie', *Revue Roumaine de géographie, 21*, 3-15.

Radalescu, N. (1972) 'Modification du milieu physico-géographique en Roumanie comme résultat de l'activité humaine', *Revue Roumaine de géographie, 16*, 1, 9-14.

—— Velcea, I. and Petrescu, N. (1968) *Geografia Agriculturii Romaniei*, Editura Stiintifica, Bucuresti.

Rogic, V. (1973) 'Regionalizacija Jugoslavije', *Geografiski Glasnik, 35*, 13-29.

Sarfalvi, B. (1965) *A mezogazdasagi nepesseg csokkenese Magyarorszagon* (Loss of agrarian population in Hungary), Akademiai Kiado, Budapest.

—— (ed.) (1966) *Geographical Types of Hungarian Agriculture*, Studies in Geography in Hungary, no. 3, Akademiai Kiado, Budapest.

Saru, Al. (1980) 'Depresiumea Transilvaniei (Regionare fizico-geografica)', *Studia Universitatis Babes-Bolyai, Geologia-Geographia, 25*, 36-48.

Sifrer, M. (1978) 'Poplavna podrocja v porecju Dravinje', *Geografski Zbornik, 17*, 7-91.

Sikos, T.T. (1977) 'Valtozatok a termeloerok teruleti elhelyezesenek gazdasagmatematikai modellezesere' (Variations for the mathematical modelling of the regional allocation of productive forces), *Foldrajzi Ertesito, 26*, 3-4, 387-402.

Somogyi, S. (1961) 'Hasank folyohalozatanak fejlodestorteneti vazlata' (The genetical pattern of the river system of Hungary), *Foldrajzi*

Kozlemenyek, 85, 1, 25-50.

Stefanescu, I. (1975) 'The migratory movement in the SR of Romania', *Revue Roumaine de géographie, 19,* 77-87.

Stehlik, O. (1981) *Vyvoj eroze pudy v CSR,* Studia Geographica no. 72, Geograficky Ustav CAV, Brno.

Straszewicz, L. (1974) *Geografia ekonomiczna europejskich krajow socjalistycznych,* Panstwowe Wydawnictwo Ekonomiczne, Warzawa.

Tufescu, V. (1972) 'Changements actuels dans la typologie des villages roumains', *Revue Roumaine de géographie, 16,* 1, 85-93.

Ujvari, I. (1980) 'Cenceptii actuale cu privire la parametrizarea factorilor fizico-geograficii ai procelesor hidrice', *Studia Universitatis Babes-Bolyai, Geologia-Geographia, 25,* 49-59.

Urbanek, J. (1968) 'Slide classification', *Geograficky Casopis, 20,* 3, 221-36.

Vresk, M. (1978) 'Gradska regija Zagreba', *Geografski Glasnik, 40,* 59-88.

Vriser, I. (1981) 'Razmestitev industrije v Jugoslaviji', *Geographica Slovenica, 12,* 5-39.

Zajicek, V. (1981) 'Funkce a vyznam experimentalnich a reprezentativnich povodi v Ceskoslovensku', *Acta Universitas Carolinae, Geographica, 16,* 2, 77-93.

Zeremski, M. (1977) 'Kriogeni procesi pleistocenske periglacialne klime', *Zbornik Radova Geografski Institut, 29,* 69-80.

Zuljic, S. (1977) 'O metodama prognoziranja rasta stanovnistva gradova', *Geografski Glasnik, 39,* 5-25.

5 THE SOVIET UNION

David J.M. Hooson

By any of the customary criteria — volume and range of published out-put, numbers of practitioners in the field, impact on the society over the past century or two and distinctiveness of approach — Russian geography would seem to deserve major attention in the international history of the subject. However, until quite recently it was a closed book for the rest of the scholarly world of geography, and was there-fore just not included in any of the general integrating works on the history and philosophy of the subject. In some ways this situation was in line with a long tradition, encouraged both from within and with-out the country, of regarding the Soviet Union as *sui generis*, a land of mystery, to be either reviled or extolled uncritically but not able to be incorporated into global studies and generalisations of universal interest. Accordingly the shock produced by that startling symbolic intrusion of the Soviet Union into the public consciousness of the rest of the world a quarter-century ago — the Sputnik — was compounded by the enigmatic, apparently almost Martian-like place from which it was launched. Conversely, the level of ignorance of the Soviet popula-tion about the outside world had been quite abysmal in Stalin's time, but was rapidly improving in a relative sense in the 'thaw' following his death.

The mutual ignorance and suspicion between Russia and 'the West' had been of long-standing. The Russians' love-hate ambivalence towards the West, and on the question of whether or not they belonged to it, had been perpetuated not only by the linguistic curtain but by a cultural isolation. Periods in which a deep-rooted inferiority complex seems to surface have alternated with bursts of Slavophile or chauvinistic expres-sions of superiority, bolstered in recent times by the presumed inevit-able rightness of the Soviet political and economic system. And the extensive borrowings of Western technological know-how, begun in earnest under Peter the Great, still continue to a substantial extent — the spectacular successes of some Soviet endeavours, such as the space programme, notwithstanding.

From the 'Western shore' the difficulties and disincentives in the way of studying the Soviet Union have been considerable a good deal of the time. Lack of precise information and of the ability of foreigners to do independent field-work, added to the bias informing most Soviet 'Marxist' presentations and explanations, have discouraged many geographers from abroad from entering the Soviet field. Some of the time it has proved difficult if not impossible to travel in the Soviet Union at all and even at the best of times large areas of the country are closed to foreigners. Some subjects are regarded as sensitive and visas may be withheld from scholars whose views and interpretations are disapproved of. On the other hand, some difficulties have been made in the United States at various times for American scholars of the Soviet Union deemed to be too sympathetic to their system.

I must say at this point that I consider myself fortunate in having first learned my Soviet geography as a teenager in Britain during the Second World War, when an enormous amount of information was perforce released and a great deal of admiration and excitement generated, which is always a powerful booster to learning. Then, after some years of disillusionment in the last years of the Stalin period, I got the opportunity to develop a Soviet specialism in the late 1950s in the United States. The next few years were exciting indeed, in both countries. The 'thaw' after the death of Stalin had released a torrent of innovation and a ferment of relatively non-conformist, reforming ideas, while travel and contacts between geographers, amongst others, of the two nations became practicable, profitable and even respectable. Following closely on these fundamental developments (accompanied by Khrushchev's repudiation of Stalin) there was the launching of Sputnik, announcing dramatically for all the world to see the accession to superpower status of the Soviet Union, a decade after it seemed to be almost destroyed. I will never forget – being a resident of Washington at the time – the electric effect of that event on the national psychology of the 'other' hitherto unchallenged superpower. The few years around 1960, in Khrushchev's Russia and Kennedy's America, were heady ones, where 'competitive co-existence' seemed, on balance, to be productive and to stimulate innovation and rethinking in both countries. Certainly it worked wonders, not only for American science in general, but also for Russian studies there. It should be noted with some sadness, however, that in retrospect these few years look like a Golden Age for the Soviet intellectual, between the Stalinist nightmare and the grey dullness and creeping repression of the Brezhnev era. It could not survive shocks like the Sino-Soviet split, the Cuban missile crisis and the Vietnam War, as well as the death of Kennedy and the ousting of Khrushchev. One has only to remember the books which were published in the USSR, of a type previously and/or subsequently banned or discouraged, and the vigorous free-wheeling arguments going on then in Soviet geography. Specifically, to compare the spirited, varied and rich flavour of a typical haul of geographical books and maps culled from the Moscow bookstores in the early 1960s with an equivalent haul in the mid-1970s is to rekindle the impression of vitality and vigour of that earlier time.

I air these personal impressions here, in the full knowledge that the interpretations of others, both in the Soviet Union and elsewhere, might well be quite different. But interpretations of such broad scope cannot fail to have a subjective element and involve the processing of multiple experiences and memories and heavy selection from a large mass of material. I presume I was asked to try my hand here at such a broad interpretation because I have, over a quarter of a century, had a more or less continuous interest in delving into the history of Russian geography from its beginnings under Peter the Great to modern times. When I began this work in the late 1950s (Hooson, 1959) there was very little material in English and the doings of Russian geography — past and present — were virtually unknown in the West. Now this situation is quite different — due in large part to the monthly *Soviet Geography: Review and Translation*, edited since 1960, indefatigably and judiciously, by Theodore Shabad; the record of what might be called the Russian School of Geography is quite well-known in the rest of the world, and is seen to compare favourably with the cumulative contribution of any other national group.

I have for some years been Chairman of an International Commission on the History of Geographical Thought, which has helped provide me with a broad comparative framework into which to fit the Russian contribution to the field. So, although I am an outsider in an obvious and literal sense, I have not only kept more or less in contact with Soviet geography and Soviet geographers over a longish period, but have considerable, if critical, admiration and sympathy for them. I must offer, however, at the outset, a disclaimer against being able to cover the whole vast field of endeavour. My aim is to try to see the wood rather than the trees, to distinguish the main directions, emphases and arguments of Soviet geography from those of other national schools of thought.

The Pre-Soviet Russian Legacy

Although this book is explicitly focused on the period since the Second World War, the developments, problems and disputations of Soviet geography in this period cannot begin to be understood properly without a more than cursory awareness of what went before, both in the pre-revolutionary period – a rich and productive time for geography – and in the 1930s, when Soviet geography was suddenly set on a diametrically different course. Paradoxically, national antecedents in geography are perhaps more crucial to the understanding of recent developments in the Soviet Union than in any other country, simply because they have provoked, and been invoked, more passionately in post-war Soviet geographical activities.

The record of Russian geography before the revolution compares favourably with that of any other country in richness, range and depth and the knowledge of this heritage, in general and in detail, forms a constant well of inspiration available to today's Soviet geographers, especially to those who feel the need of reform. But in the rest of the world this history of Russian geography is still mainly a closed book because of its lack of adequate incorporation and integration into the history of geographical thought as a whole, though this situation has certainly improved in the last decade. Hence the continuing need here to attempt to outline the more distinctive themes, phases and personalities which make up the world of Russian geographical ideas.

The foundations were laid in the early eighteenth century under Peter the Great with far-reaching surveys of the rapidly expanding Empire – its physical features and resource potentials in particular. This was the period of the first and most explosive of the bursts of modernisation which have characterised much of Russian history down to the present. Then, as now, the introduction of Western know-how was a prime stimulus and this period saw the founding of the Academy of Sciences – still very powerful in the Soviet Union – and various universities. While foreigners were crucial in these beginning stages, a home-grown contingent of original Russian scholars grew up as the century progressed, from Kirilov (Novlyanskaya, 1958), who produced the first Russian atlas, to Krasheninnikov (1764), who revealed the fascinating natural and human features of the far-off Kamchatka peninsula to the world. Geography arguably became the most successful intellectual vehicle through which Russian scientific achievements were developed and displayed at this time (Vucinich, 1963, p. 60). Building on these exhilarating foundations, firmly laid down not only by the personal enthusiasm of Peter the Great but also by the general popular appeal which was generated, geography took a somewhat different turn in the early nineteenth century, under the impact of revolutionary events and fermenting literature from Europe and eventually America. The emphasis was increasingly placed on economic and social considerations, with the explicit intention of creating a workable foundation for social reform and the improvements of the lot of the peasants. The mixture of altruism and patriotism, alongside a pervading love for the Russian landscape and environment, led a quite distinctive blend of geography to take root at that time.

The most important institution to bring together these various ways of thought in the mid-nineteenth century was the Imperial Russian Geographical Society. Its rapid and many-sided growth testified to how pressing was the need for it in the national life. It quickly became the most successful and popular, in the best sense, of the Russian learned societies. It initially acted as an umbrella for studies in anthropology, archaeology, geology and meteorology as well, and this range was matched by that of the political and philosophical stances of the major protagonists in the new geographical enterprise — old-style explorers, social reformers, scientists and humanists. The reforming spirit of the 1860s, which saw the long-awaited emancipation of the serfs, gave a renewed sense of purpose to the integration of this highly varied group of adopted geographers in spirit. Peter Semënov Tian-Shansky, who took the helm of the Society in this period, in some ways embodied this integration in his own person. He had studied in Berlin with Ritter and, at the same time as he was carrying out his basic explorations in the Tien Shan mountains, he was active in social reform movements, especially the committee for the emancipation of the serfs. His work on land ownership and regionalisation actually brought forth praise from Karl Marx (Saushkin, 1953), while at the same time he built up and maintained a secure and honoured place in the Tsarist Russian establishment. HIs temperament and catholic tastes enabled him to bring together diverse elements and infuse them with the common purpose of promoting geography, as a scientific, social, patriotic and even moral endeavour. The fact that he was the leading light in the Geographical Society for the last half-century of Tsarist rule was very fortunate, for this turned out to be the Golden Age of Russian geography which has been the inspiration — explicit or otherwise — for recent Soviet geographers (Hooson, 1968).

It should be recalled that this was a period in the Russian empire of suddenly accelerating rates of economic growth, renewed exploration and colonial conquest in Asia, as well as an American-style settlement of the open spaces, railway building on a continental scale and expansion of exports, especially of wheat. Russia was thus coming to share much in the way of the experience of both America and the West European imperial powers, although of course it diverged markedly in its political and social forms. In spite of the inevitable restrictions of the Tsarist autocracy, the spirit and momentum of change and ferment led to an intellectual, scientific and educational revival in which several original geographers played an important part. Among the most prominent of these scholars who were the most influential for Soviet geographers was A.I. Voeikov, a wide-ranging, internationally confident scientist and humanist, who became a world authority on climatology and the heat and water balance concept in particular. At the same time he was engaged in studies of population and climatic analogues of the Russian Empire across the world as a framework for guiding agricultural innovations and experiments, especially in the more southerly lands which were coming under Russian rule. In a period when much was being made of the influence of nature on man, Voeikov systematically studied and publicised the record of human impact on the earth and thus, along with George Perkins Marsh in America, he may be regarded as an important forerunner of the modern spirit of conservation. He did a great deal to advise and promote, for example, measures for arresting soil erosion, stream control and shelter belts (Voeikov, 1963).

Complementing Voeikov's work was that of another original scientist, Dokuchaev, who achieved international recognition, particularly because of his basic theory of soil formation and classification, with climate recognised as the critical factor. He demonstrated that integrated approach to the study of the interacting processes of the natural environment, and the focus on bio-climatic 'natural zones', rather than geomorphic features, which has set the dominant tone of Soviet physical geography to this day. His classic study was on the *Russian Blackearth (Chernozem)* (Dokuchaev, 1948) and he was always mindful of the connections with agricultural practice and reform, implications for rational land use and conservation measures. The continuing importance of landscape science (*landshaftovedenie*), which is basically the composite study of these natural zones and which has little counterpart in the rest of the world, owes much to Dokuchaev's initial scheme.

Another humane and liberal scholar, D.N. Anuchin, who more than anyone else was instrumental in institutionalising geography as an independent discipline in the university, should be mentioned here. He became head of the joint department (*Kafedra*) of geography and ethnography at Moscow University in 1884 and exercised a powerful influence on both subjects, in universities, schools and popular education, for nearly 40 years. His scholarly work ranged from geomorphology to ethnography and his stated focus was on human-centred regional geography. He founded a remarkable geographical journal (*Zemlevedenie*) in 1894, which was fully comparable with its French contemporary the *Annales de géographie* in scope, volume and quality, and he found time to write a large number of articles, reviews and notes for it, steering it and embodying its spirit until his death in 1923 (Yesakov, 1955).

Thus in the mid-1880s, when Anuchin was inaugurating geography at Moscow University, Voeikov was publishing his monumental *Climates of the World* and Dokuchaev his *Russian Blackearth*, and the Geographical Society under Semënov Tian-Shansky's direction was pursuing its many-sided activities and popular appeal, the condition of Russian geography seemed at least as vigorous and original as that of the leading countries of Western Europe. Together with the lively interest of famous scientists like Mendeleev and Vernadsky, humanists like Sinitsky and Lamansky and even Marxists like Plekhanov (1934), this flowering in the last three or four decades of the Tsarist era laid down rich and distinctive foundations and a corpus of themes, methods and philosophies which have been of high importance to geographers of the Soviet period.

This is the basic justification I advance for what may seem undue attention paid here to the pre-Soviet period. For the paradox is that, in spite of the revolutionary overturning of Russian society after 1917, the underlying continuity of thought in Russian geography over the last hundred years has been as marked as that of any other national school. Some of the more enduring features of this tradition may be summarised here: (1) a functional and integrated way of looking at the natural environment, with an emphasis on bio-climatic rather than geomorphological features: physico-geographical concepts, such as that of Voeikov's heat and water balance and Dokuchaev's natural zonation, still have key significance today in Soviet geography; (2) a greater attention to the human impact on nature than vice-versa and a concern for ameliorative, conservationist schemes, especially for agriculture; (3) a preoccupation with 'regionalisation', usually with 'applied' ends; (4) concern for a natural resources inventory, and national atlases, accompanied by that combination of patriotism and philanthropy, of ambivalent pride and inferiority *vis-à-vis* 'the West', which remains a persistent characteristic of many Russian intellectuals to this day; and (5) a regional, integrated approach to geography, combining natural and human phenomena, including the landscape as well as regional schools.

Although, as we shall see, many of these traditional guiding precepts were severely stultified in the Stalin period, their roots proved strong enough to survive it, and also to prove surprisingly compatible with Marxist thought in the broad sense.

Revolution, Continuity and Radical Change

When considering the impact of the advent of the Soviet regime on the progress of geography, certain contrasting periods should be recognised. The first decade or so after the Bolshevik revolution saw a considerable degree of continuity in intellectual life and institutions, and even in the economy, though it was interspersed with social and political experiments and minor upheavals. It has to be remembered that this period was wracked with civil war, foreign interventions and famine. By 1928 the *per capita* income of the country was still at about the level it had been in 1913, if not lower, and four-fifths of the population were still peasants and villagers.

Although most of the prominent geographers, noted above, died just before the revolution, the field was dominated in the 1920s by their students, who kept alive the methods, aims and philosophies of their mentors, which happened to be by no means necessarily antithetical to general Marxist principles. Seminal, even monumental, works like Kruber's (1923) *General Earth Science*, Vernadsky's (1929) *The Biosphere*, V.P. Semënov's (1928) *Region and Country* (see also Kaganskii, Polyan and Rodoman, 1980) and Berg's (1950) *Natural Regions of the USSR*, came out in this period, reflecting faithfully the spirit of pre-Soviet Russian geography.

However, the intellectual climate, in geography, as in all other aspects of Soviet life, had changed abruptly by 1930, with the imposition of Stalin's first Five-Year Plan. Geographers, like others, were fully mobilised and *cadres* of specialists were quickly trained and sent off on immediate practical tasks. Very quickly, this had the effect of seriously fragmenting geography, so that the traditional broad, integrated philosophy was sidestepped. Alongside this, partly because so many of the practical projects involved inventories of physical resources and partly because work in human or economic geography became politically risky (prominent geographers were arrested for using western location theory to advise against projects which were dear to Stalin's heart, for instance: V.V. Pokshishevskii, in Baransky, Nikitin, Pokshishevskii and Saushkin, 1965, p. 482), the make-up of the geographical profession became steadily more exclusively physical in emphasis. Thus the physical geographers *ipso facto* became dominant in the key professional institutions, in particular the Institute of Geography of the Academy of Sciences, whose directors are vested with very considerable powers over trends, policies and the fortunes of the geographical profession and of the lives of individual geographers in the country. Under their direction, physical geography attained a privileged and dominant place, squeezing economic or human geography almost out of existence, or into the not altogether open arms of economics. Even more significant, and destructive to some of the basic traditions of Russian geography, was the promulgation of a doctrine of strict philosophical separation between physical and human geography, based on the proposition that fundamentally different laws underlay the development of the phenomena in the 'two geographies' (Grigoriev, 1951, Gerasimov, 1955, 1956). The necessary corollary was that integrated studies of physical and human phenomena and their relationships were ruled philosophically and scientifically 'illegal' and were regarded as leading inevitably to 'loaded' interpretations associated with 'vulgar geographism' or environmental determinism, which was still supposed to characterise Western geographical practice, and which had, routinely and officially, been branded as 'bourgeois', 'imperialistic', etc. in a vituperative and xenophobic way, especially at the outset of the Cold War. Unfortunately one of the worst examples of these tracts, by Academician Grigoriev, was the only Soviet work cited by Hartshorne (1959).

Physical Bias in the Early 1950s

A perusal of the geographical literature in the early 1950s reflected the cumulative constraints and tendencies of the Stalin period described above. The papers were overwhelmingly physical in orientation and even these had become much more specialised and often, surprisingly, divorced not only from relevance to economic planning but also showing signs of having appreciably moved away from the landscape, natural zonation or heat-water balance principles of Dokuchaev and Voeikov (Grigoriev, 1951). There was a good deal of attention to paleogeography, or evolutionary geomorphology, and to specialised studies in glaciology, hydrology and biogeography. However 'the Great Stalin

Plan for the Transformation of Nature', launched in the late 1940s, predictably led to a rash of rather hurriedly prepared and sycophantic articles on such topics as forest shelter-belts and the diversion of rivers, lauding the notion of the 'conquest of nature' (Gerasimov, 1951). There were also, inevitably, articles in praise of Stalin's geographical thoughts along these and other lines.

Some work did appear on the history of geography, mostly concerned either with biographies of Russian geographers or with accounts of Russian exploration. Apart from Berg (1962), these did not touch upon controversial matters of theory or judgement, but seemed to indicate the resurgence of acceptable Russian nationalism in the Great Patriotic War of 1941-5 on the one hand and the desire to avoid involvement in topics which might become, as they had been before, vulnerable politically. Very little attention was paid to foreign countries except for diatribes against 'imperialism' or 'geopolitics', and articles on the new 'People's Democracies' of Eastern Europe. Papers on economic or human topics were very few and far between in the regular geographical periodicals, although an exception must be made for those in the series *Voprosy Geografii* (Questions of Geography), an irregularly issued but vigorous journal launched in 1946 by N.N. Baransky and other Moscow geographers. In general the official picture of Soviet Geography in the early 1950s, as appearing for instance in an authoritative article in the London *Geographical Journal* by the then director of the Academy of Sciences' Institute of Geography (Grigoriev, 1955), is one of a predominantly physical collection of sciences, oddly apolitical, parochial and unexciting. About the same time a new edition of the *Great Soviet Encyclopaedia* (1957) was being prepared, whose geographical editor was the same Academician Grigoriev, which depicted geography as exclusively physical, with a detailed description of work in its specialisms. Economic geography, by contrast, was accorded only one sentence, submerged in a tract on 'Economic Science' (Saushkin, 1958). There had in fact been good work done in economic and population geography, in scattered ways and places and in spite of official discouragement, but the power of the official Geographical establishment, and the cumulative, generally bleak, results of the Stalin period, were plain to see in the mid-1950s and provided a powerful pretext for the remarkable intellectual ferment and reactions of the next phase.

Grand Disputation and Renewal in the Years About 1960

In retrospect, the period between the mid-1950s and the mid-1960s stands out as *the* watershed in the history of geographical thought and practice in the Soviet Union since the Second World War. Roughly the same period would be said by many people also to constitute the watershed in American geography, but it may be worthwhile at this point to suggest certain fundamental differences in what was going on in the two newly-emerged superpowers at that time. In the mid-1950s, when Soviet geography, as we have seen, had become overwhelmingly physical and 'legally' and doctrinally split into two (unequal) parts, American geography had the opposite problem, having become dominantly human in orientation and yet holding to the traditional philosophical axiom of the essential unity of geography as a whole. Both national schools of geography were poised for a radical restructuring and reorientation of their assumptions, theories and practice in the mid-1950s, accompanied by bitter arguments and a propensity to polarisation. However, whereas the American revolution focused on techniques and methodological issues, and was not in any sense harking back to earlier national traditions, the Soviet contemporaneous equivalent was more fundamentally philosophical, much less technical, and contained a very important element of intellectual nostalgia for a broken national heritage in Russian geography.

It may be helpful here to recall the general atmosphere in the Soviet Union in the mid-1950s to appreciate the changes under way in the geographical profession. All aspects of Soviet intellectual life in the years following the death of Stalin exhibited a rapid thaw, or the release of a ferment of reforming ideas and vigorous argumentation which had lain muted for a quarter-century. These were, for Soviet geography, fluid, and even heady, times, during which, for example, a vigorous new periodical *Geografiia i Khoziaistvo* (Geography and the Economy) was launched (in 1958). The volume of new publications grew quickly – well illustrated, lively and often controversial – and the balance shifted suddenly, in relative terms, from the physical to the economic side of the subject. International contacts were beginning to be eagerly re-established – 1956, for example, saw Soviet geographers attending their first International Geographical Congress for two decades, and also reviewing at great length the volume *American Geography: Inventory and Prospect*, in the first issue of the new 'Geographical Series' of the Academy of Sciences (1956). The change in tone exemplified by this review – keenly interested and not unsympathetic – and by other reviews of foreign work, contrasted strikingly with the vituperative equivalents which had been common form a few years earlier. This was the beginning of the Khrushchev decade, which can now be looked back upon, in spite of its lurching inconsistencies and 'hare-brained schemes' (as they were afterwards dubbed), as something of a Golden Age for the Soviet intellectual. Books by previously (and subsequently) banned authors were published, there was a renewal of a relatively enlightened nationalism and internationalism, 'de-Stalinisation' and a quite radical attempt, even if ultimately unsuccessful, to restructure the economy on less centralised lines. The national ebullience and new-found confidence was boosted immeasurably by the launching of Sputnik and, before the setbacks of the growing Sino-Soviet rift and the Cuban missile crisis, a relatively open, outgoing and relaxed foreign policy. The atmosphere in academic circles was more conducive to innovation, reform and the challenging of official doctrines than it had been at any time since the 1920s, although of course the power of the institutional and party establishment had also become far more entrenched.

The Anuchin Affair

Among the many challengers to the geographical *status quo* in this period of ferment, the most articulate, uncompromising and even charismatic was V.A. Anuchin, who was then teaching at Moscow University. The arguments and challenges came to a head most forcefully and inescapably in Anuchin's book *Theoretical Problems of Geography*, which was published in 1960 (1977). This was the first book in Soviet history to set out to investigate the theoretical basis of geography as a whole through historical and philosophical analysis, while coming to definite conclusions about what the subject is and where it should go from here. As not only a scholarly but also a hard-hitting book, aimed at demolishing the prevailing doctrine of duality in Soviet geographical theory and practice, it was bound to be opposed by the geographical establishment, notably in the Academy of Sciences' Institute of Geography, by then, as now, headed by Academician Gerasimov. In fact, the relatively tolerant political atmosphere of the time notwithstanding, it is doubtful whether Anuchin's book would have seen the light of day if it had not previously found one or two influential supporters among the senior geographers — the most important · of whom was N.N. Baransky. Luckily for Anuchin, Baransky was then, at the age of 80, revered and loved by the profession at large and, being originally a Siberian revolutionary and even a friend of Lenin, as well as a humane, objective scholar, his authority was unassailable. However it is indicative that, even with the wholehearted, energetic support of Baransky and his influential successor at Moscow University, Yu. G. Saushkin, amongst others, Anuchin was inevitably to have an uphill fight on his hands to make his case (Anuchin, 1962). One should have a proper feeling for the deadly seriousness of challenges to official doctrine by academic juniors in the Soviet Union (a problem by no means unknown in the West either), the courage needed, the height of the stakes and the unpredictable nature of the political climate within which the arguments had to be conducted. The fact that the heads of the two most influential institutions, the Institute of Geography of the Academy of Sciences and the All-Union Geographical Society, came out early and unequivocally against Anuchin and the heresy of a 'unified' view of geography, would have daunted most would-be challengers. But the combination of the relatively heady spirit of innovation and tolerance in the air around 1960, the vigorous longevity and moral authority of Baransky, the intellectual toughness and persistence of Anuchin, and the hitherto submerged yearning for the recapture of a national heritage, proved strong enough to overturn the existing conceptual structure of geographical theory and practice. But even so it was an intellectual and power struggle of titanic proportions, comparable in Soviet terms with the more notorious Lysenko controversy in genetics and biology in general.

The Battle-Royal in the Early 1960s

The astonishingly mixed reception to Anuchin's book in the few years following its publication had many facets to it (Hooson, 1962). The intellectual arguments centred on the philosophical rightness of a unified approach to physical and economic geography or one of strict separation between two fields of study with fundamentally different 'laws'. Apart from the purely philosophical or scientific planes, both sides, of course, were at pains to establish their approach as more squarely in line with Marxist principles and the Marxist-Leninist texts than their opponents', with some inevitably strained connections advanced. Then there was the invocation of the letter and the spirit of the pre-Stalin Russian geographers. Finally there was the reference to the practical needs of Soviet planning and the avoidance of disasters and unnecessary costs to the economy and the environment. Although it would be wrong to imply that there were not some telling arguments from both sides, it became clear as the debate progressed that the monists were increasingly winning 'on points' over the dualists in all

these three crucial planes. However the existing academic power-struc-
ture, as well as the persistence of some of the methods of 'guilt by association' and character-assassination used under Stalin's 'cult of the
personality', were revealed along two further avenues taken during the
controversy. The first was demonstrated by the fact that Anuchin was
turned down unanimously when he presented his book for the Doc-
torate of Geographical Sciences at Leningrad University in 1961. The
following year he submitted it to Moscow University where the public
dissertation defense turned out to be a dramatic affair attended by
hundreds of people. By that time the case had become a *cause célèbre*
and rank-and-file sympathy on his home ground was strong. But even
there the final vote fell just short of the required two-thirds majority,
reflecting the importance of essentially external votes.

The other factor, equally powerful, involved the time-honoured use
of innuendo by identifying the views of Anuchin and his Soviet supporters with those of American or other Western geographers and *ipso
facto* tarring them with the brush of guilt by association with allegedly
'bourgeois' or 'reactionary' views. It seems that in all of these cases,
which can be so damaging in the Soviet milieu, the victims were
Anuchin and the other proponents of a more unified geography.
Saushkin, for instance, although a powerful figure as the head of economic geography in Moscow, was severely pilloried by a group of economists and geographers who, in a collective letter to a Soviet journal,
charged him with 'distortion' and 'traitorous' statements in an article
that he had published in the American Journal *Economic Geography*.
In his hard-hitting reply, Saushkin gave no quarter or concession,
deploring 'the fact that some geographers have still not got rid of the
rough methods of unsubstantiated accusations and intimidations used
in the period of Stalin's personality cult' (Saushkin, 1963).

The Turning of the Tide

When such a prolonged, bitter and apparently fundamental disputation
had been allowed to rage in the scientific literature for several years, it
became necessary for the Communist Party to step in, as had been customary in such cases. Thus, in what proved to be the last year of the
Khrushchev regime, a leading ideological spokesman, L.F. Ilyichev
(1964), in a statement before the Presidium of the Academy of
Sciences, denounced the Stalinist definition of the environment as 'a
purely natural category' — laid down 30 years before, and the fact
that this edict had become the pretext for the construction of 'an insur-
mountable wall' between nature and society, with deleterious effects on

the Soviet economy and planning processes. In the context of the disposition of authority in the USSR, this new pronouncement heralded a setback for the established theoretical doctrine of 'legal separation' of physical and economic geography.

Further emboldened by these developments, Anuchin thenceforth broadened the arena of the controversy to include not only non-geographical academic periodicals, notably in philosophy, but also various national, semi-popular, journals and magazines. The most highly visible of these forays was a series of six articles in *Literaturnaya Gazeta* (1965), initiated by Anuchin and rounded off by Saushkin, with the explicit support of the editors. Although the series was by no means devoid of disagreement, the eventual result was to censure strongly the general policies and priorities of the Institute of Geography of the Academy of Sciences and its director, Gerasimov, particularly the lack of provision for economic and integrated regional studies and the undue dominance of physical geography. (In fact Gerasimov was severely criticised by some of his own staff and also by his Academy Praesidium and urged to shift the emphasis radically towards more integrated, economically-oriented studies.) In his winding-up remarks, Saushkin made a proposal, with the blessing of the editors, that the economic geographers should be 'the conductors of the geographical orchestra', i.e. basically guiding the research priorities in the physical branches as well. This would have involved a drastic reorientation of the whole structure and practice not only of the Institute but the universities as well – which would have been truly revolutionary.

In a remarkably short time the official geographical pronouncements were brought into line with the new directives. For instance, in his programmatic statement of 1966, 'The Past and the Future of Geography', Gerasimov (1966) proposed supplanting the traditional topical specialities of the subject with a set of integrated synthetic topics focused on specific problems and regions, with a markedly strengthened economic component. Then in 1968 he presented this new look to the International Geographical Congress in India as typifying 'the Soviet approach to geography'. Later he followed this up, along with other leaders of the profession, by further elaboration of the theory of 'constructive geography', as it came to be called.

Thus the official line and accepted doctrine in Soviet geography in the late 1960s had become radically different from that of the late 1950s, so much so that it seems in retrospect to qualify as a genuine revolution, propelled from below and eventually blessed from above, for good practical as well as theoretical reasons, consonant not only with pre-Stalin Russian, but also Marxist, traditions.

Recent Changes

What, then, is the nature of Soviet geography in the early 1980s, nearly two decades after the unprecedented ferment and theoretical restructuring of the mid-1960s? As in so many other aspects of life in the Soviet Union since the end of the Khrushchev era, the answer to this apparently simple question is difficult to arrive at in clear, unequivocal terms, fraught as the situation has been with contradictions, paradoxes and nuances. Inevitably the state and development of individual disciplines, and that of Letters and Sciences in general, could not fail to reflect the spirit and tenor of the times, essentially the Brezhnev years, which have been characterised by declining growth and increasing repression of independent minds. Compared with the fermenting dozen years or so which followed the death of Stalin, the subsequent years have witnessed the onset of a duller, less controversial, more comfortable (for some) and sluggish atmosphere, which has been particularly demoralising for young people of initiative and originality, and has led to a disturbing increase in cynicism, opportunism and corruption in Soviet society as a whole. Although lip service to Marxist-Leninist principles is still paid as a matter of course, sincere personal conviction about them is now rarely encountered. Not only are the spontaneity and excitement of the arguments (and of the books produced) around 1960 noticeably absent today but there is considerably less basic information, for example, statistical and cartographic, available than there was then. Salishchev (1976), the chief Soviet academic cartographer, has voiced his disappointment and frustration at the slow progress and narrow range of production of thematic maps and atlases in the past decade or so.

Not surprisingly, then, an aura of dullness, lack of innovation, and even a colourless scholasticism have come to permeate a great deal of the quite voluminous geographical literature of the last decade at least. New ideas and methods, and above all controversy and zest, which abounded in the literature of the Khrushchev period and its immediate aftermath, are rare now, though not of course entirely absent. It seems as if, after the dust of the ideological revolution and subsequent reformation had settled, the situation reverted to a more normal one for the Soviet Union, with such discussion as there was directed to means rather than ends. However, some general features of today's geographical literature may be mentioned, impressionistic though such comments must be.

First, compared with most Western geographical periodicals, a typical issue of any of the chief Soviet ones, namely the organs of the Academy of Sciences' Institute of Geography, the All Union Geographical Society and Moscow University's Geographical Faculty respectively, still contains a heavy representation – usually a majority – of articles in physical geography, though the proportion is much smaller than it was three decades ago. The persistence of the assumption that geography is still to be regarded as mainly physical geography is clearly reflected in the authoritative and scholarly history of geographical ideas by Isachenko (1971) which, while considerably more 'humane' than Grigoriev's official pronouncements of two decades earlier, is still much less so than any equivalent survey in the West and incidentally probably less than would satisfy Marx, Engels, Plekhanov or Lenin. Within the physical segment, geomorphology is still the best represented branch, although more applied topics have been incorporated. The Russian traditions of landscape science and natural zonation have been revived since Berg's death in 1950, in spite of having been discouraged by Grigoriev. Characteristically, there has been a strong reassertion of this framework of thought, including the regional and applied approaches, in physical geography in connection with the celebrations of the recent centenary of Berg's death (Markov, 1977; Murzayev, 1977a). Murzayev's excellent, monumental physical geography of Middle Asia (1958) set a new tone for the genre by starting with a concise discussion of the economic, political and population patterns of the region, and its general geographical context. The then novel realisation that the criteria for the selection and presentation of material were functional was shown by no means to exclude the study of processes or even a genetic approach to the understanding of the interacting elements of the natural environment. Murzayev is still contributing actively to the promotion of a judicious mixture of continuity and change in Soviet physical geography (1977b).

Moreover the world-wide ecology-awareness movement in the 1970s has naturally helped to preserve a breadth of vision and purpose in Soviet physical geography today. Two other recent reassertions of interest, which go some way towards integrating Soviet physical geography, perhaps irrevocably, with the human world, have had much attention recently. One is the study of anthropogenic landscapes or, broadly, the 'Influences of Man on the Landscape', which is the title of a recent special issue of *Voprosy Geografii* (Kovalev, ed., 1976). The 23 articles in it examine the humanised landscape from a wide variety of

angles, from the historical and economic to the engineering and geo-
chemical. The other major new integrating theme is known as nature
management (*prirodopolzovaniya*), which is clearly in tune with the
world-wide conservation spirit of the 1970s (Pryde, 1972, 1983). It is
closely tied in with the planning process, as illustrated by the new book
by Anuchin (1978), who is a leader of the Council for the Study of
Productive Forces, an influential planning agency. Economists and
engineers have often been accused, as in other countries, of ignoring
the environment as well as people, and this movement is proving to
have growing protective power. Other recent statements on the physical
elements of nature management (Gvozdetskii *et al.*, 1976), on its equiv-
alence with historical geography (Zubakov, 1978), resource renewal
(Gerasimov, Armand and Yefron, eds., 1971) and the global impact
(Ryabchikov, 1975), reinforce the pervading nature of this movement
and its congruence with its counterparts in other countries. It appears
also that Soviet geography is beginning to exhibit a more balanced and
less inhibited view of 'environmental influences' on human activities
than is often the case in the West (Matley, 1982; Chappell, 1975; Spate,
1963) and Anuchin's campaign against 'indeterminism', amongst others
(Anuchin, 1977, 1978), may be counted a success here. Popular books
on environment and resources (Lvovich, 1963; Armand, 1964) and
analyses of particular equations, such as water and farming (Lvovich,
Koronkevich and Yurevich, 1979), northern oil exploitation (Denisova,
1976), and the forests along the Baikal-Amur Mainline (BAM) railway
route (Gorovoy, 1978), are increasingly becoming more comprehensive.
It should be noted here, however, that fully three-quarters of the
departments (*Kafedry*) of the Geographical Faculty at Moscow
University are devoted to physical geography (including glaciology,
oceanology, hydrology, etc.), including about two-thirds of the teaching staff. Since this Geographical Faculty is by far the most prestigious
of its kind in the Soviet Union (as well as, incidentally, probably the
world's largest teaching and research institution in geography), and although it was the centre of agitation for a more integrated geography
in the 1960s, the resistance to change in the actual structure seems
clear.

'Economic geography', long the only acceptable designation for non-
physical aspects of the subject in Soviet parlance, has also been expanded, 'naturalised' and humanised, with more attention to social and
demographic phenomena (Kovalev, 1963, 1980; Pokshishevskii in
Yefremov, Kolosova and Sdasink, eds., 1981, pp. 50-9), even tourism
(Yefremov, 1975), as well as more traditional topics such as 'terri-

torial-production complexes' (Bykov, 1976). Although there was more than a flurry of interest in following up Western quantitative-spatial-model-building precept and practice in the late 1960s and early 1970s (Saushkin, 1971), it appears that interest has dwindled, as indeed it has elsewhere. Recent revisionist reviews of works like Harvey's *Explanation in Geography* (Saushkin, 1975) and Bunge's *Theoretical Geography* (Gerenchuk, 1980), books which had been enthusiastically translated into Russian, reflect some of these changes of heart. Interest in a largely urban-based human ecological approach, again following certain American exemplars, has been mainly directed by Medvedkov (1974, 1979), who headed a separate human ecology unit in the Academy of Sciences' Institute of Geography. However at this time of writing he and his wife (Medvedkova, 1980) are suffering prolonged persecution in that Institute and are being refused permission to emigrate. An interesting recent departure has been a book on the economic geography of the world oceans (Markov, ed., 1979). It covers a wide range of topics, from transport and fisheries to settlement and energy, with better map illustrations and a more international approach than are usual in Soviet publications.

There is still very little of what would be recognised as cultural, historical or political geography elsewhere, and some of the few extant examples, such as the eccentric pieces by Gumilev on 'ethnogenesis' (1973, 1981), have come in for considerable criticism (Drozdov, 1977). Human geography remains rather heavily economic in emphasis, which is not surprising given general Marxist precepts, in spite of a relative recent humanising trend within that accepted framework. Given the importance accorded to cities as centres of revolution and innovation in Marx's teachings (as the counterpoise to the 'imbecility of rural life'), one might expect the regular geographical periodicals to be full of urban studies, whereas in fact they rarely occupy more than 5 per cent of a typical issue. However there is *in toto* a large corpus of such studies from a wide variety of sources, as has been amply demonstrated, e.g. by Harris (1970), Khorev (1968, 1971), Listengurt (1975).

This same judgement (about the relative paucity of representation) could in some ways still also apply to regional geography, in spite of the implied green light it received from the reformation of the 1960s, but of late, as in the West, this old staple and, to some, crowning chapter in the history of geography seems to have been staging a serious come-back in the Soviet Union. In particular, a recent important straw in the wind was a whole issue of *Voprosy Geografii* (Yefremov *et al.*, 1981) devoted to regional geography. This semi-regular periodical,

'Questions of Geography', founded in 1945 by Baransky, has published an average of three to five issues every year on a wide variety of topics, natural and human, and integrated them around problems of real significance. I would therefore exempt this treasure, of over 120 volumes, to date, from the strictures I applied to the regular journals – in fact I would hold this excellent series up as the brightest single star in the firmament of Soviet geography, clear proof of its potential and resilience. The volume on regional geography was explicitly dedicated to Baransky, who can be regarded as the greatest Soviet exponent of the regional school (Baransky, 1980), which had such deep roots among the pre-Soviet Russian geographers. The collection of 20 articles covers a wide range of attitudes to the subject, and the whole is infused with a conviction, not voiced for a long time with such unison, that regional geography is once again to be regarded as the most distinctive and thoroughly geographical of the various branches of the field, and the least susceptible to being co-opted by other sciences.

A recent book, *Soviet Geography Today: Aspects of Theory* (Lavrov, 1981) published in English in Moscow, can also be commended to non-Russian-speaking geographers who want to have in one compass a sampling of recent writings by 13 notable Soviet geographers of divergent views and backgrounds. Their articles appeared earlier in the Soviet literature and have been well translated. Most of them make quite lively reading, notably those by K.K. Markov, a geomorphologist who has become a forceful protagonist for the unified approach to geography, A. Isachenko, a leading historian of geography, and opponent of a unified geography, who discusses determinism and indeterminism among foreign geographers in a generally refreshingly, undogmatic way, and Saushkin, whose wide-ranging piece on 'Projects for the development of Geography' ends the collection. The average age of the contributors is high (in fact five have died since the book was published) but it includes the original Moscow geographer Boris Rodoman, who writes, with Gokhman, on trends in theoretical geography. The editor, S.B. Lavrov, Vice-President of the Geographical Society, has brought together writers who had been sharply opposed to each other, such as Gerasimov and Anuchin, and Kalesnik and Saushkin. In a preface he explains the laudable aim of disseminating the work of a varied group of Soviet geographers to their Western colleagues, and mentions, among other things, the 'generally optimistic view of the development of geographical science' and 'the recognition of the dominance of integrating tendencies in the system of geographical sciences' (Lavrov, 1981, pp. 5-16).

In early 1983 another significant straw in the wind appeared in the form of an article in the official Party organ *Kommunist*, by the same Lavrov, together with Anuchin and Agafonov (Agafonov, Anuchin and Lavrov, 1983). Once again, this is a strong plea for academic geography to range itself more wholeheartedly and effectively behind integrated studies of nature and economy, as well as regional studies, and subjects the Institute of Geography of the Academy of Sciences, and its director, Gerasimov, to hard-hitting criticism for failing to do this adequately so far. This is an unusually authoritative vehicle for such an urgent, categorical and forceful statement and, in the context of the Soviet system, almost takes on the character of a command. But, impressive as it seems to be, one cannot avoid a feeling of *déjà vu*, for the same charges, levelled unequivocally at the very same institution and person, were made in the official press nearly two decades ago. The director of the key Institute, approaching 80 years old, the call for whose replacement was clearly implied at that time, is still in full charge there, with additional all-powerful, *ex officio* posts such as Chairman of the National Committee of Soviet Geographers, head of delegation to all international conferences, including the power of final sanction over whether individuals can attend, and related powers. There is nothing peculiar to geography in this, of course, since the inflexibility of the institutional structure laid down in the early Stalin era, and the gerontocracy it seems inevitably to engender, is widely recognised as a major and intractable obstacle to reform and progress in the Soviet Union. Even Andropov's apparently earnest attempts at fundamental economic reforms in 1983 seem likely to founder in part upon this rock, and since the general socio-historical context is crucial to the progress and character of geographical thought and practice, as all forms of endeavour, we must recognise the difficulties in the way of releasing the undoubted energies, ideas and aspirations of the bright younger generation of Soviet geographers — heirs to a great tradition.

Prospects

I must confess to a certain feeling of disappointment, tinged however with a mitigated optimism, as a concerned and sympathetic observer of Soviet geography for over a quarter-century. I did, after all, introduce to the English-speaking world, before that now indispensable American journal, *Soviet Geography – Review and Translation*, had been born, the story of the unprecedented intellectual ferment released in Soviet geography in the years following the death of Stalin (Hooson, 1959). Later, I made the first survey in the West of the extraordinary scholarly legacy of pre-Soviet Russian geography (Hooson, 1968), which had hitherto been almost completely absent from the world literature on the history of geography. In both of these reports I found no difficulty in warming to the intellectual excitement and the resurrection of this legacy in tune with the times, and the emergence of a surprisingly fruitful marriage between Russian traditions and Marxist imperatives, to produce a composite picture of the continuity of a truly distinctive Russian school, which compares favourably with the other better recognised national schools of the world.

I still feel that the basic strengths are there — awareness of a heritage which could meet the needs of a composite subject which can both provide for a geographically literate general public (unfortunately lacking in the United States at present) and for its innovative and fruitful involvement in the social and economic planning and natural preservation of the country. If the renewed calls for more truly integrated studies, the strengthening of the regional approach with the additions of infusion of the historical approach to the organisation of society, as Annenkov has recently advocated (Annenkov, 1980), were implemented, then Soviet geography could yet be the equal of that of any country in the world. Its strength and breadth in physical geography, compared with most other countries, are great. With a deeper human, historical and regional component, it could be an example to us all, an example sorely needed when geography's fortunes in several important nations (including notably the United States) are unaccountably clouded. I can sympathise somewhat with Gerasimov's frustration (1983), in a review of a Russian edition of Haggett's textbook, that, still today, very few Russian geographers are mentioned in the general Western literature as seminal contributors in the history of the field. In part this relative omission is surely unjustified, but in part it is also understandable, and not only because of the language barrier. Given the full engagement and encouragement of the latent originality, imagination and industry of the younger, and of many of the older, generation of Soviet geographers, along lines now being strongly advocated there (e.g. Agafonov, Anuchin and Lavrov, 1983) and the reawakening of pride in their Russian heritage and the potentials of fruitful planning, there should be no problem about proper recognition, at home and abroad. And we in the rest of the world of geography, beset, as many of us are, with our own problems of image and role, would be deeply grateful.

References

The range of possibilities is endless, of course, and very many worthwhile publications inevitably cannot appear here. For those who do not read Russian, attention should be drawn to a rich continuing resource, the monthly journal *Soviet Geography: Review and Translation*, edited by Theodore Shabad in New York, which has been publishing selected articles from the Soviet geographical literature since 1960. A perusal of these volumes can provide an accurate 'feel' for the nature and trends in geography in the Soviet Union, although it also reflects, to some degree, the interests and needs of the world-wide English-readership it exists to serve. Many of the references here are from this journal and, in the interests of space, the citation of the original Russian title and place is omitted. This information is, of course, to be found in *Soviet Geography*. Only articles which have not been translated into English, which includes almost all of those published before 1960, are cited in the original Russian. For space reasons also, *Soviet Geography: Review and Translation*, is referred to in the listing as simply *SG*.

Abramov, L.S. and Murzayev, E.M. (1981) 'Vklad N.N. Baranskogo v razvitie Stranovedeniya S.S.S.R.' in Yefremov, Yu. K., Kolosova, Yu. A. and Sdasink, G.V. (eds.) (1981) *Stranovedenie: Sostoianie i zadachi, Voprosy Geografii, 116*, 14-27.

Agafonov, N.T., Anuchin, V.A. and Lavrov, S.B. (1983) 'The Present Tasks of Soviet Geography', *SG*, June, 411-22.

Alampiyev, P.M. *et al.* (1963) 'Letter of Protest Regarding Yu. G. Saushkin's Article in *Economic Geography*, 1962, No. 1', *SG*, January, 60-2.

Annenkov, V. (1980) 'Razmyshlenya o geografii vosmidesyatikh', *Yezhegodnik Zemlya i Liudi*, Moscow.

Anuchin, V.A. (1962) 'On the Criticism of the Unity of Geography', with foreword by N.N. Baransky, *SG*, September.

—— (1972) *Teoreticheskie Osnovy Gegraffi*, Moscow.

—— (1977) *Theoretical Problems of Geography* (English edition edited by R. Fuchs and G. Demko), Ohio University Press, Columbus.

—— (1978) *Osnovy Prirodo-Polzovaniya: teoreticheskii aspekt*, Moscow.

—— (1982) *Geograficheskii faktor v razvitii obshchestva*, Moscow.

Armand, D. (1964) *Nam i Vnukam*, Moscow.

Baransky, N.N., Nikitin, N.P., Pokshishevskii, V.V. and Saushkin, Yu. G. (eds.) (1965) *Ekonomicheskaya Geografiya v SSSR*, Moscow.

——(1980) *Izbrannye Trudy* (collected works), with commentary by V.A. Anuchin, 2, vols., Moscow.

Berg, L.S. (1950) *Natural Regions of the U.S.S.R.* (English edition edited by J.A. Morrison and C.C. Nikiforoff) New York.

——(1962) *Istoriya Russkikh Geograficheskikh otkrytii*, Moscow.

Budyko, M.I., Vinnikov, K. Ya., Drozdov, O.A. and Yefimova, N.A. (1979) 'Impending Climatic Change', *SG*, September, 395-411.

Bykov, V. (1976) *Bratsko-Taishetskii Promyshlenny Raion*, Irkutsk.

Chappell, J.E. (1965) 'Marxism and Geography', *Problems of Communism, 14*, 12-22.

——(1975) 'The Ecological Dimension: Russian and American Views', *Annals of the Association of American Geographers, 65*, 144-62.

Demko, G. (1965) 'Trends in Soviet Geography', *Survey, 55*, 163-70.

Denisova, T.B. (1976) 'Types and stages of economic development in new pioneering regions of the northern USSR', *SG, 17*, 679-90.

Dokuchaev, V.V. (1948) *Russkii Chernozem* in *Izbrannyi Sochineniia, 1*, Moscow.

Donde, Alexander (1983) 'Theoretical Problems of Soviet *Stranovedeniye'*, *SG*, March, 188-203.

Drozdov, O.A. (1977) 'A Critique of L.N. Gumilev's Work on Ethnogeography at a Meeting at Leningrad University', *SG*, February, 119-28.

French, R.A. (1961) 'Geography and Geographers in the Soviet Union', *Geographical Journal, 127*, 163-4.

Gerasimov, I.P. (1951) 'Stalinskii plan preobrazovanya priorody i uchastie geografov, v ego osyshchestvlenii', *Izvestiia Akademii Nauk SSSR, Seriia Geograficheskaia, 5*, 5-15.

——(1955) 'Sostoianie i zadachi sovietskoi geografii na sovremennom etape ee razvitiia', *Izvestiia Akademii Nauk SSSR, Seriia Geograficheskaia, 3*, 14-15.

——(1956) 'Le rôle de la géographie dans la construction socialiste en URSS et les tendances actuelles de son évolution', *Essais de Géographie;* recueil des articles pour le XVIII Congrès International Géographique, Moscow-Leningrad.

—— Ignat'yev, G.M., Kalesnik, S.V., Konstantinov, O.A., Murzayev, E.M. and Salishchev, K.A. (eds.) (1962) (English edition, C.D. Harris (ed.)) *Soviet Geography: Accomplishments and Tasks*, American Geographical Society, New York.

——(1966) 'The Past and the Future of Geography', *SG*, September, 3-14.

—— Armand, D.L. and Yefron, K.M. (eds.) (1971) (English edition, W.A.D. Jackson (ed.)) *Natural Resources of the Soviet Union: Their Use and Renewal*, W.H. Freeman, San Francisco.

—— (ed.) (1976) *A Short History of Geographical Science in the Soviet Union*, Progress Publishers, Moscow.

—— (1983) 'Contemporary Anthropogeography at the Chorological Level, Pragmatism, Quantitative Techniques and the Trend Toward Constructivism (with reference to Peter Haggett's *Geography: A Modern Synthesis'*, *SG*, April, 267-79.

Gerenchuk, K.I. (1980) 'On Theoretical Geography', *SG*, January, 42-7.

Gokhman, V.M., Medvedkov, Yu V. and Preobrazhensky, V.S. (eds.) (1976) *Novye Idei v geografii: (1) Problemy Modelirovaniya i informatsii*, Moscow.

—— and Rodoman, B. (1981) 'Certain Trends of Development in Theoretical Geography in the USSR' in S.B. Lavrov (ed.), *Soviet Geography Today : Aspects of Theory*, Progress Publishers, Moscow, 187-200.

Gorovoy, V.L. and Shlykov, V.M. (1978) 'Basic Trends in the Development of the Forest Industry along the Baykal-Amur Mainline', *SG*, February, 84-98.

Grigoriev, A.A. (1951) 'O nekotorikh voprosakh fizicheskoi geografii,' *Voprosy Filosofii, 1*, 193.

—— (1955) 'The State of Soviet Geography', *Geographical Journal, 121*, 429-39.

Gumilev, L.N. (1968) 'On the Anthropogenic Factor in Landscape Formation', *SG*, September, 590-602.

—— (1973) 'The Nature of Ethnic Wholeness', *SG*, September, 467-75.

—— (1981) 'The Epoch of the Battle of Kulikovo', *SG*, February, 112-9.

Gvozdetskii, N.A., Yefremov, Yu. K., Isachenko, A.G., Kogay, N.A., Preobrazhenskii, V.S. and Ukleba, D.S. (1976) 'Physical-Geographical Fundamentals of Nature Management', *SG*, May, 291-302.

Harris, C.D. (1970) *Cities of the Soviet Union*, Rand McNally, Chicago.

Hartshorne, R. (1959) *Perspective on the Nature of Geography*, Rand McNally, Chicago.

Hooson, D. (1959) 'Some Recent Developments in the Content and Theory of Soviet Geography', *Annals, Association of American Geographers, 49*, 73-82.

—— (1962) 'Methodological Clashes in Moscow', *Annals, Association of American Geographers, 52*, 469-75.

—— (1968) 'The Development of Geography of Pre-Soviet Russia', *Annals, Association of American Geographers, 58*, 250-72.

—— (1977) 'Introduction to the English Edition', in V.A. Anuchin, *Theoretical Problems of Geography*, Ohio University Press, Columbus.

Ilyichev, L.F. (1964) 'Remarks About a Unified Geography', *SG*, April, 32-4.

Isachenko, A.G. (1971) *Razvitie Geograficheskikh idei*, Mysl, Moscow.

—— (1972) 'On the Unity of Geography', *SG*, April, 195-219.

—— (1981) 'Determinism and Indeterminism in the Work of Foreign Geographers' in S.B. Lavrov (ed.), *Soviet Geography Today*, 201-41.

Kaganskiy, V.L., Polyan, P.M. and Rodoman, B.B. (1980) 'V.P. Semënov-Tyan-Shanskii's *Region and County: Its Present-Day Relevance and Meaning*', *SG*, June, 355-63.

Khorev, B.S. (1968) *Gorodskie poselenia S.S.S.R.*, Moscow.

—— (1971, 2nd edition 1975) *Problemy gorodov*, Moscow.

Kovalev, S.A. (1963) *Selskoe Rassellnie: geograficheskoe issledovanie*, Moscow.

—— (ed.) (1976) *Vozdeistvie cheloveka na landshaftu, Voprosy Geografii, 106.*

—— and Koval'skaya, N. Ya. (1980) *Geografiya naseleniya*, Moscow.

Krasheninnikov, S.P. (1764) *The History of Kamtschatka and the Kurilski Islands, with the countries adjacent*, London (printed by R. Raites for T. Jefferys, geographer to His Majesty).

Kruber, A.A. (1923) *Obshchee Zemlevedenie*, Moscow.

Lavrishchev, A. (1969) *Economic Geography of the U.S.S.R.*, Progress Publishers, Moscow.

Lavrov, S.B. (ed.) (1981) *Soviet Geography Today: Aspects of Theory*, Progress Publishers, Moscow.

—— and Sdasiuk, G.B. (1981) 'Globalnye problemy i sovremmenoe stranovedenie' in Yefremov *et al.* (1981), 60-8.

Listengurt, F.M. (1975) 'Criteria for Delineating Large Urban Agglomerations in the U.S.S.R.', *SG*, November, 559-68.

Literaturnaya Gazeta (1965). Articles by nine geographers beginning with 'A Sad Tale About Geography', and ending with 'The Today and Tomorrow of Geography', *SG*, September, pp. 27-56.

Lvovich, M.I. (1963) *Chelovek i Vody*, Moscow.

—— Koronkevich, N.I. and Yurevich, R.A. (1979) 'The Water Aspect of the Geography of Agriculture in the USSR', *SG*, November, 515-33.

Markov, K.K. (1977) 'Recollections About Lev Semyonovich Berg (on the centennial of his birth)', *SG*, January, 19-22.

—— (ed.) (1979) *Ekonomicheskaya Geografiya Mirovogo Okeana*, Leningrad.

—— (1981) 'General Aspects of Geography' in S.B. Lavrov (ed.), *Soviet Geography Today*, 71-113.

Matley, I.M. (1966) 'The Marxist Approach to the Geographical Environment', *Annals, Association of American Geographers, 56*, 97-111.

—— (1982) 'Nature and Society: The Continuing Soviet Debate', *Progress in Human Geography*, September, 367-96.

Medvedkov, Yu. V. (1974) 'Ekologicheskie problemy bolshogo goroda i puti ikh resheniya', *Voprosy Geografii, 96*, 32-42.

—— (1979) 'Geografy ob ekologii cheloveka' in Medvedkov (ed.) *Novye idei v geografii: (4) Geograficheskie aspekty ekologii cheloveka*, 378-93.

Medvedkova, O.L. (1980) 'Components in the Evolution of Urban Systems: Theory and Empirical Testing', *SG*, January, 15-29.

Murzayev, E.M. (1958) *Sredniaia Aziia – Fiziko-geograficheskaia kharakteristica*, Moscow.

——(1977a) 'The Career of L.S. Berg', *SG*, January, pp. 4-12.

—— (1977b) 'Things that are almost but not quite well known in physical geography', *SG*, May, 287-301.

Novlyanskaya, M.G. (1958) *I.K. Kirilov i ego Atlas Vserossiskoi Imperii*, Moscow.

Plekhanov, G. (1934) *Fundamental Problems of Marxism*, M. Lawrence, London. (translation of 1928 Russian edition)

Pokshishevskii, B.B. (1951) *Zaselenie Sibiri – Istoriko-geograficheskoe ocherki*, Irkutsk.

Pokshishevskii, V.V. (1965) 'Sergei Vladimirovich Bernstein-Kogan' in Baransky *et al.* (eds.) *Ekonomicheskaia Geografiya v SSSR*, 482.

—— (1966) 'Relationships and Contacts between Prerevolutionary Russian and Soviet Geography and Foreign Geography', *SG*, November, 56-76.

—— Mints, A.A. and Konstantinov, O.A. (1971) 'On New Directions in the Development of Soviet Economic Geography', *SG*, September, 403-16.

—— (1979) 'Soviet Economic Geography: Six Decades of Development and Contemporary Problems', *SG*, March, 131-9.

Pryde, P.R. (1972) *Conservation in the Soviet Union*, Cambridge University Press, Cambridge.

—— (1983) 'The Decade of the Environment in the U.S.S.R.', *Science, 220*, 274-9.

Rodoman, B.B. (1968) 'The Organized Anthroposphere', *SG*, November, 784-96.

Ryabchikov, A.M. (1975) 'Problems of the Natural Environment in Their Global Impact', *SG*, June, 402-12.

Salishchev, K.A. and Aslanikashvili, A.F. (1976) 'The Present State and Tasks of Complex Mapping in the USSR', *SG*, May, 303-13.

Saushkin, Yu. G. (1953) 'Rabota Karla Marksa nad Russkogo Geografa P.P. Semënova Tian-Shanskogo', *Voprosy Geografii, 31*.

—— (1958) 'Kishcheznovenniu geografii i ekonomicheskoi geografii v tome S.S.S.R. Bolshoi Sovietskoi entsiklopedii', *Izvestia Vsesoiuznogo Geograficheskogo Obshchestva, 90*, 190-1.

—— (1963) 'Yu. G. Saushkin's reply to a letter of protest by nine Soviet geographers', *SG*, October, 25-30.

—— (1971) 'Results and Prospects of the Use of Mathematical Methods in Economic Geography', *SG*, September, 416-27.

—— (1975) 'Review of D. Harvey's *Explanation in Geography*', *SG*, October, pp. 538-46.

—— (1980) *Economic Geography: Theory and Methods*, Progress Publishers, Moscow.

—— (1981) 'Prospects for the Development of Geography' in S.B. Lavrov (ed.), *Soviet Geography Today*, 247-62.

—— (1982) 'The Lessons of Nikolay Nikolayevich Baransky, *SG*, December, 736-43.

Semënov Tyan-Shanskii, V. (1928) *Raion i Strana*, Moscow.

Spate, O.H.K. (1963) 'Theory and Practice in Soviet Geography', *Australian Geographical Studies, 1*, 18-30.

Vernadsky, V.I. (1929) *La Biosphère*, F. Alcan, Paris.

Voeikov, A.I. (1963) *Vozdeistvie cheloveka na prirodu*, with introduction by V.V. Pokshishevskii, Moscow.

Vucinich, A. (1963, 1970) *Science and Russian Culture, Vol. 1: to 1860, Vol. 2: 1860-1917*, Stanford University Press, Stanford.

Yefremov, Yu. K. (1969) 'The Landscape Sphere and the Geographical Environment', *SG*, May, pp. 248-56.

—— (1975) 'Geography and Tourism', *SG*, April, 205-16.

—— Kolosova, Yu. A. and Sdasiuk, G.V. (eds.) (1981) *Stranovedenie: Sostoianie i zadachi, Voprosy Geografii, 116*.

Yesakov, V.A. (1955) *D.N. Anuchin i Sozdanie Russkoi Universitetskoi Geograficheskoi Shkoly*, Moscow.

Zubakov, V.A. (1978) 'On the Content and Tasks of Historical Geography (as the History of Nature Management)', *SG*, March, 170-9.

Zvonkova, T.V. and Saushkin, Yu. G. (1977) 'Interaction between Physical and Economic Geography', *SG*, April, 245-50.

6 THE UNITED KINGDOM

R.J. Johnston and S. Gregory

Geography in the United Kingdom is a substantial pedagogic and academic discipline, but it lacks any professional status and has no explicit non-educational role. It is widely taught in the schools, at all levels, and in the early 1980s some 5,400 students annually entered the universities, polytechnics and colleges of higher and further education to read for degrees either exclusively in, or containing a substantial component of, geography. On graduation, they proceeded to a wide variety of careers in only one of which — secondary and tertiary education — are they likely to retain their identity as geographers.

Research in geography is undertaken in the tertiary teaching institutions, especially the universities. There are no geographical research institutes in either the public or the private sector. Thus most active researchers — excluding postgraduate students — are also university/polytechnic/college teachers; they combine the functions of advancing and propagating knowledge. As researchers, most of them operate as individuals, pursuing topics of personal interest with little resource aid (although library, computing and technical support is generally good and is freely available as part of the institutional infrastructure). Large research teams are rare, and no 'research schools' dominated by one or a few scholars characterise British geography (Freeman, 1979). Indeed, because in most institutions staffing policy for geography departments has been to cover a wide range of topical and regional specialisms — in order to provide a full course for undergraduate students — there is little clustering of researchers with similar interests. Further, there is relatively little concentration of postgraduate studentships (now only 70-80 new ones per annum for the whole of geography).

Research in geography is concentrated in the university teaching departments, therefore. The staff, most of whom have tenure for life, combine teaching and research, in both of which they have considerable discretion over what they do. Thus, any major 'movements' or 'trends' in geographical scholarship reflect individual choice rather than central direction, although that choice is environmentally influenced (Stoddart, 1981a). Despite this autonomy, most of the major changes have been stimulated by a small number of innovators, whose lead has then been followed, in a general if not a specific sense. As the discipline has expanded in recent decades, however, so the potential choice and the number of would-be innovators has grown. The result is a vital discipline, with a very wide spread of scholarly interests.

The Institutional Context

The academic traditions of British geography can be traced to the early nineteenth century, but the development of the discipline has been a twentieth-century phenomenon (Freeman, 1961, 1980a, 1980b). In the first two decades, lectureships were established at ten universities, and some individuals were already employed to teach geography in another department (e.g. geology). The first Honours Degree course was established at Liverpool in 1917; in the next seven years, Aberystwyth, the London School of Economics, University College London, Cambridge, Manchester and Sheffield established similar (exclusively geography) degrees. By 1945 there were 25 university departments of geography, the situation having changed but little since 1930 (Stoddart, 1981b).

Most of these departments were small, with an academic staff exceeding four being unusual. Teaching loads were heavy, and individuals were required to contribute courses on subjects well outside their research interests. A considerable breadth of outlook was part of the disciplinary ethos, and educational objectives (especially preparing graduates for careers as school teachers) provided the main departmental *raison d'être*. Facilities and resources were few; staff shared offices; there might be a single teaching room; a class library was a luxury; and collections of maps were only slowly accumulated. There were no custom-built cartographic and physical laboratories, and field teaching depended on visual interpretation rather than careful measurement.

Despite these constraints, the volume of research was considerable, with a few individuals being particularly active. The influence of Wooldrige, Steers, Linton, Vaughan Lewis and Austin Miller in geomorphology and of Crowe, Garnett and Austin Miller in climatology was counterbalanced by Fleure, Bowen, Estyn Evans, Forde and Unstead in social geography and Rodwell Jones, Stamp, Beaver, Buchanan and Smith in economic geography. These 'broadly-based specialists' provided the foundation for disciplinary expansion after the Second World War.

The war itself had a substantial impact on the later events in geography. First-hand experience, often in adverse conditions, of distant parts of the world stimulated concern with a range of social and economic problems to be tackled by 'area studies' specialists, while the employment of geography graduates on the government's *Admiralty Handbooks* provided integrated data sources far superior to anything previously available. Skills in the use and interpretation of aerial photographs, competence in surveying and map-making, and the introduction of synoptic meteorological concepts into descriptive climatology, were added to the geographers' stock-in-trade as a result of the war effort, and a concern with planning and post-war recovery was extended, following pioneer work by Stamp, Fawcett, Gilbert and others.

Building on these foundations needed institutional growth, which took place in the 1950s. A few new departments of geography were established but, more importantly, those already existing expanded substantially – in both students and staff. Further expansion took place in the 1960s, with several departments having between 20 and 30 academic staff and graduating 100 or more students annually. Many obtained new facilities, including several custom-built departments with properly-designed laboratories, and resources, including support staff.

Several new university departments were opened in the 1960s and 1970s, along with 25 in the polytechnics (which were created in 1966); the *raison d'être* and structure of the latter were closely matched to the university pattern. Also in the 1970s several Colleges of Education and other further education institutions began to offer geography degrees (many of them offering geography as part of a combined degree). This boom in tertiary-level provision was fed by an expansion of geography teaching in the schools, for most incoming undergraduates have studied the subject continuously from the age of 11 until leaving school at 18.

This continuous growth for 80 years has been fostered by the Geographical Association, established in 1892 to encourage and support teachers. The Association operates nationally and through its local branches. Its periodicals (*The Geography Teacher*, 1901-26; *Geography*, 1927- ; *Teaching Geography*, 1975-), occasional publications, central lending library, annual conferences, field excursions and a wide range of local activities, plus representations to government and other bodies on educational issues, have done much for the advancement of the discipline, at all levels. For research, the Institute of British Geographers was founded by 74 members in 1933, largely as a 'publishing co-operative'. It now has some 1,700 members, produces two journals (*Transactions*, 1935- ; *Area*, 1969-), holds a large annual conference, and contains 17 study groups which organise a variety of conferences and other activities. And alongside is the senior society, the Royal Geographical Society, founded in 1830. This has a long tradition of support for exploration (Cameron, 1980), and it has been closely involved with the educational role of geography, especially at the country's older universities (Freeman, 1980a). There are regional bodies, too, of which the Royal Scottish Geographical Society is the premier.

Support for research (including postgraduate studentships) comes from two research councils – the Natural Environment Research Council, which covers physical geography, and the Social Science Research Council, for human geography and area studies. The number of grants and studentships awarded each year is small, however.

The Precursors of Change, 1945-60

In the first 15 years after the Second World War, the pre-war attitudes to and characteristics of academic geography were largely re-established (Wooldridge and East, 1951). The major metamorphosis came in the 1960s, but slow modifications in the previous decade provided the foundation for that change.

Central to the pre-war attitudes was the almost universal acceptance of regional synthesis as the apogee of geographical work (Johnston, 1983c). This was reflected in courses and texts on regional geography that contributed much systematised and pre-digested information about the physical landscape and human activity, but rarely provided an integrated synthesis. British geographers frequently castigated themselves for failing to produce regional works comparable to those of the French Vidalian school and for not evoking the personality of different regions. Through the 1950s this criticism was extended, with an increasing number of geographers questioning the value and utility of the regional synthesis, and the links between environment and human activity (as in the continuing debate over environmental determinism: Martin, 1951; Jones, 1956; Spate, 1957) were of little interest to a substantial proportion of younger researchers.

The decline of interest in regional geography was associated with – as both cause and effect – a growth of specialist, systematic interests, traditional and new. Geomorphology was by far the most popular among physical geographers, with the model developed by W.M. Davis, and extended by Baulig (1935) and King (1953), providing the major stimulus to research; within Britain, Wooldridge and Linton (1939; see Jones, 1981) were the research leaders. The paradigm encapsulated in the phrase 'structure, process and stage' dominated, with close links to geology (especially stratigraphy) providing the structural focus but with virtually no attention to processes. The evolution of landforms through sequences of uplift and erosion, interrupted by glaciations, provided the main intellectual issues (Brown and Waters, 1974).

New fields of study for geomorphologists were being opened up, however, such as climatic geomorphology (stimulated by Cotton, 1947, and Peltier, 1950). The spatial variations in processes associated with different climates were recognised as important influences on landform genesis, and new literature, new environments and new problems were posed to those geomorphologists who chose to work in Australasia, tropical Africa and North America. A concern with deep weathering, the weathering front, laterites, pediplanation and inselberg formation in tropical areas led to work on chemical weathering (for example Thomas, 1972), thereby introducing a new focus and methodology to geomorphological work.

Climatological interests were extended, too, with the introduction of synoptic climatology (Frisby and Green, 1949) and of weather-type terminology (Lamb, 1950) by those with wartime weather forecasting experience. It became accepted that understanding the physical processes in the atmosphere was essential to climatological work, and the concept of the water balance – developed in the physical sciences (e.g. Penman, 1948) – was central to this work. The relatively weak background in the physical sciences of most geographers, plus the absence of a major hydrological component to geographical work and little acknowledgement of the needs for numeracy in climatological/hydrological research, inhibited advances in this area at that time, however.

Biogeography and soil study were extreme minority interests in the 1950s. Russian, climate-based, work provided the standard approach to mapping global soil patterns, but there was very little geographical research on soils except among a few geomorphologists who were employed as soil surveyors (in Britain and notably in tropical Africa), and on the part of agricultural geographers (Buchanan, 1935). Similarly, there was little research in biogeography. The concepts of the ecosystem and of vegetational climaxes were slowly introduced to teaching syllabuses; the main research advances were associated with the growing geographical concern with pollen analysis (Oldfield, 1960).

In all aspects of physical geography there was an increased emphasis during this period on field data collection — it was the era of learning geography through the soles of one's feet. Erosion surfaces were mapped (Balchin, 1952; Brown, 1957); techniques of morphological mapping were devised (Waters, 1958); local climates were measured; and soils were augered. But there was little laboratory analysis, however, for facilities were normally unavailable. Research reports were mainly descriptive interpretations of mapped phenomena, therefore, with no statistical analysis or mathematical modelling.

Within traditional human geography, it was the fields of historical and land-use (especially agricultural) study that flourished. With the former, the work of Darby stimulated analysis of past periods through mapping and interpretation (again, largely descriptive) of archival material (Darby, ed., 1936, 1977), such as that in the Domesday Book and nineteenth-century censuses. With the latter, the influence of Stamp's (1947) land-use survey was extensive. Land-use inventories were promoted as necessary bases for planning, and geographers were presented as uniquely able to provide these bases, through synthesising land-use data with those on the physical environment (Freeman, 1958).

The newer developments in human geography at this time concerned the study of towns and of industries. In urban geography, there were concerns with identifying central-place hierarchies (Smailes, 1946; Dickinson, 1947), with the definition of urban hinterlands (Green, 1950), and with understanding the genesis of urban morphology (Conzen, 1960; Whitehand, ed., 1982). The first texts were published (Smailes, 1953), but the seminal contribution was undoubtedly Jones's (1960) analysis of the social geography of Belfast which combined his early work at Aberystwyth in rural social anthropology with that of the Chicago urban ecologists. In industrial geography, the work of Smith (1952), Rawstron (1958) and Estall and Buchanan (1961) moved away from description of particular places and establishments towards an understanding of the principles underlying location decisions. Political geography focused on boundary studies and the study of the state as a region (East and Moodie, eds., 1956).

By 1960 British geography was substantially different from its structure a decade earlier. There had been no revolutions, but rather a slow evolution away from the regional synthesis, for which systematic work provided necessary inputs, towards systematic analyses as ends in themselves. Nevertheless, with hindsight it is possible to identify, in the growing disenchantment with the subjective elements of regional description and landform interpretation, the shift towards detailed data collection, and the increasing concern with spatial pattern in both physical and human geography, the precursors of the major shifts in the decade that followed. At the turn of the decade, several major regional texts, and new editions of standard texts, were published (e.g. C.A. Fisher, 1964; W.B. Fisher, 1950; Cole, 1961; Spete, 1954), but these were the last substantial outputs of a declining field of study.

The Watershed Years: Quantification Rampant

The 1960s formed a decade of methodological fervour in British geography. At the beginning, very few undergraduates were introduced to statistical procedures other than simple averages; ten years later, very few received no courses in such methods, and a considerable number were introduced to sophisticated multivariate techniques, to mathematical modelling and to computer use. Associated with this shift in teaching was a marked change in the content and style of the research output. The 'quantitative revolution' was far from popular with the disciplinary establishment (Stamp, 1966; Taylor, 1976), but it was widely adopted among the younger generation who were filling the new posts in the expanding departments. By the early 1970s their impact was being felt in the school syllabuses too.

Quantification is a tool that can be used in a variety of ways and philosophies. It was as a descriptive methodology that it was introduced to British geographers, from two sources. The first was endogenous, and was associated with the work of climatologists (e.g. Crowe, 1936) — the geographers with the longest tradition of numerical analysis. The main impact came with Gregory's (1963) *Statistical Methods and the Geographer*, which was an introductory student text (see Gregory, 1976, for a résumé of this impact). The exogenous source was North America, where Burton (1963) claimed that the quantitative revolution was over by the early 1960s. Several British geographers who had studied there — such as Chorley, Garner and Haggett — returned excited by the methodological ferment, the results of which were filling the American journals.

Initially quantification was associated with precision in description. Soon, however, it was linked with a philosophical shift among geographers (itself slow and implicit only) towards the positivist conception of science. In this, the goal is the development of generalisations (laws) through the testing of scientific hypotheses, with such generalisations having predictive powers. From a discipline that had concentrated on the description and interpretation of the unique, geography was being rapidly transformed into one that sought to generalise, and which used statistics inferentially to generate its law-like statements.

In this new, generalising geography the systematic branches of the discipline rapidly achieved dominance of both research agenda and teaching curricula. The work undertaken was influenced by a small number of seminal books and, in human geography especially, by a particular approach to modelling. Two of the books were edited collections, based on courses which explicated the 'new geography'. *Frontiers in Geographical Teaching* (Chorley and Haggett, eds., 1965) provided the first contact for many with the positivist goal − it included, for example, a major attack on Davisian geomorphology with its avoidance of process study. The highly successful *Models in Geography* (Chorley and Haggett, eds., 1967) provided more substantial reviews covering nearly all systematic fields within geography; a large, much of it non-geographical, literature was summarised and its relevance to geographical studies of the spatial organisation of society and of process-form relationships in physical geography was made clear. The book was both bible and agenda for future work.

The third book, Haggett's (1965) *Locational Analysis in Human Geography*, presented not only an implicit philosophy (positivism) and an explicit methodology (hypothesis-testing via inferential statistics) but also a substantive, holistic framework for human geography. Its focus was spatial form (frequently termed spatial structure), which Haggett divided into five components: nodes, hierarchies, networks, flows and surfaces. (A sixth − diffusion − was added in the second edition: Haggett, Cliff and Frey, 1977.) The book integrated and reviewed a great volume of substantive research, much of it in other disciplines. Its thesis was that human geographers should seek generalisations about the spatial organisation of society; within that, the role of distance as an influence on aspects of economic, social and political decision-making was of paramount interest.

The last of the important books published in the 1960s was the first to make the philosophy of positivism explicit, to both human and physical geographers. Harvey's (1969) *Explanation in Geography* illustrated 'the fantastic power of the scientific method', which had been implicit but unrecognised during the 'quantitative revolution' (p. vii). Explanation is the goal of science, which presents all events as the consequences of cause-effect sequences that illustrate the operation of one or more scientific laws. Harvey outlined in considerable detail the components of scientific method, its languages, and its potential for geographical research.

The 'new geography' of the 1960s concentrated on the analysis of spatial patterns, and is frequently termed 'spatial science'. Human and physical geographers shared the goal of quantitative description of and scientific explanation for mapped patterns and their interrelationships, although the models from which their hypotheses were drawn came from different systematic disciplines. They shared the excitements of developing techniques for description and analysis, as illustrated by Haggett and Chorley's (1969) *Network Models in Geography*, and continually demonstrated since in the activities of the Quantitative Methods Study Group of the Institute of British Geographers (IBG) (Wrigley and Bennett, eds., 1981; Gregory, 1983).

During the 1970s, as discussed below, there has been a parting of the ways between human and physical geographers and, among some human geographers, some disillusion with the goals of positivism. Nevertheless, there has been much work in the spatial science genre; a large proportion of research studies use quantitative methods and (at least implicitly) are set in the generalising framework of positivism. Almost every geography undergraduate in the early 1980s takes compulsory courses in quantitative analysis and computer use – as increasingly do 16-18-year-olds in schools, whose (university-inspired) syllabuses are dominated by the 'new geography' of the 1960s.

The closest links between human and physical geographers at present involve quantitative work. Among both, for example, there has been adoption of the *systems approach* introduced by Chorley (1964; Chorley and Kennedy, 1971). This provides a holistic perspective (as initially demonstrated in statements regarding the ecosystem framework: Stoddart, 1965, 1967) within which physical-human interrelationships are stressed (Bennett and Chorley, 1979; Huggett, 1980), and provides a framework for mathematical modelling of environmental and spatial systems (Thomas and Huggett, 1980), within which some innovative work has been done (Wilson, 1970, 1981a, 1981b).

A second area of continued interaction between physical and human geographers has been the field of *spatial statistics*. In the early quantitative work in the 'new geography', conventional statistical techniques were applied with little realisation that their use in the analysis of spatial patterns faced major inferential problems. Work at Bristol, stimulated by Haggett, explored these issues of spatial autocorrelation and developed methods for circumventing the problem of the interdependence of observations because of their spatial proximity (Cliff and Ord, 1973, 1981). Work on the technical issues has been advanced substantially, with sophisticated investigations of interrelationships in both time and space, including diffusion patterns (Bennett, 1979; Cliff *et al.*, 1975, 1981).

From Form to Process

Quantification, and the philosophy of science with which most geographers implicitly associated it, entered the discipline in Britain as a major force during the 1960s. By the end of the decade the research activities of both physical and human geography were dominated by this approach, and adherents of the 'new geography' were brought together by shared interests in methodology and techniques. Since then, although the pace of technical developments and the substantive research output have by no means faltered, the link between the two parts of the discipline has been dissolved somewhat. This discussion of the 1970s must treat the two separately.

Physical Geography – Incorporation with the Natural Sciences

The major shift in physical geography has been a move away from a primary concern with spatial forms, or patterns, *per se* and towards analysis of the processes, or mechanisms, that create those forms. This has involved a greater input of material from the natural sciences, a continued employment of the positivist philosophy and methodology, and, consequent on a realisation of the complex interdependencies involved in landform genesis, a greater use of the systems approach (e.g. Trudgill, 1977a). Study of processes in operation requires sampling to provide needed data – assumed to represent the generality of the processes – and the validity of postulated mechanisms is assessed via statistical procedures, notably in climatology, hydrology, pedology and ecology (Unwin, 1977). The product of a piece of physical geographical research is rarely now a map of erosion surfaces, of vegetation mosaics, of soil types or of climatic averages but rather an attempted account of processes operating within the physical environment to produce these landscape components; that account probably draws on known physical and/or biological laws, expressed as mathematical models.

This shift was heralded by Chorley (1962) in the early 1960s, initiating the impact of Americans such as Leopold and Langbein (1962), Scheidigger and Langbein (1966), Schumm and Lichty (1965) and Strahler (1964) – all of whom have had a much greater impact on British than on American geography. In geomorphology, changes in the subject were influenced by work in hydrology (Ward, 1967) and hydraulic engineering (Leopold, Wolman and Miller, 1964; Gregory and Walling, 1973), and by a wider interest in engineering matters, as in the study of slope form and development (Carson and Kirby, 1972; Young, 1972). The focus was on the operation of contemporary processes, especially measurement of their rates, with the latter leading to a return to a broader, evolutionary viewpoint (Brunsden and Thornes, 1979). A consequence of this shift in interest has been the greater potential contribution of physical geographers to the solution of current environmental problems, as illustrated by the volume of consultancy work undertaken by geomorphologists (Brunsden, Doornkamp and Jones, 1978; Briggs, 1981); to some, this indicates the need for development of geomorphology as a separate professional discipline (Worsley, 1979).

This interest in processes has stimulated greater attention to pedology and the role of the soil in environmental systems (Curtis, Courtney and Trudgill, 1976). Similarly, there has been a growth of interest in biogeography (Tivy, 1971; Watts, 1971; Pears, 1977) following earlier publications by Eyre (1963) and Edwards (1964). In all of this work on environmental systems, knowledge of physical and chemical laws is vital, while the acquisition of skills in laboratory experimentation and analysis has been essential. Since the mid-1960s university departments have been able to acquire at least some of the needed equipment, and so underwrite the costs of this more 'scientific' work. There has been less growth of work in climatology, however, despite its recognised relevance to many problems of food production and urban living; major limitations are undoubtedly the relative lack of competence in atmospheric physics and applied mathematics, plus the high costs of data collection. Several texts have been provided covering the relevant processes (e.g. Barry and Perry, 1973; Atkinson, 1981a, ed., 1981b) and their climatic impact (Lockwood, 1974; Chandler and Gregory, 1976), and the potential of using remotely-sensed imagery for dynamic and descriptive climatology has been explored (Barrett, 1967, 1974).

The availability of remotely-sensed data has had a considerable influence on much physical geography in recent years (Barrett and Hamilton, 1982). Its use has required expansion of various technological facilities, including computing, but geographers have been able to contribute substantially to developments in the use of these data sources, especially with regard to tests of their ground truth. Associated with this work have been developments in automated cartography and computer graphics (Rhind and Adams, 1980).

Although the systems approach suggests a holistic physical geography, in fact the 1970s has seen the development of substantially separate research fields, as illustrated by three new journals — *Earth Surface Processes and Landforms, Journal of Biogeography* and *Journal of Climatology*. Thus recent years have seen a diversification and intensification of work in physical geography, coupled with an awareness of the interrelationships among the various parts and of the applied potential of much research.

Human Geography – from Spatial Science to Humanism and Marxism

The quantitative revolution introduced a new way of studying spatial forms by human geographers. The search was for generalisations about the spatial organisation of society and in the 1960s this was set in the context of rational decision-making, with regard to profit-maximising in economic location decisions (hence the popularity of the works of Christaller, Lösch, von Thünen and Weber) and to social welfare in residential choice (following the lead of the Chicago school of urban ecology). Hypotheses were derived from the models and were tested against mapped patterns — although criteria for verification were generally vague.

Testing these hypotheses produced at best only limited success; the models did not fit, because, it was suggested, of their unreal representation of human decision-making. So inductive approaches were substituted, and the goal was pursued through attempts to generalise about the reality of decision-making, especially with regard to the influence of distance. Positivism remained the guiding philosophy for surveys of why factories were sited where they were, why people used particular shopping centres, what influenced migrations, how people organised their views of space, and so on (see Gold, 1980). Thus, it was claimed, a form of spatial determinism was replaced by a more *human* geography that focused on decision-making. To some, however, the positivist goal of generalising about human behaviour denied the singularity and humanity of each individual. A more *humanistic* geography was demanded.

This reaction to positivism involved the revival of previous approaches, especially in historical (which has flourished in recent decades) and cultural geography, and an explanation of their philosophical underpinnings (which had not been undertaken before). The goal was empathetic understanding of man-in-environment, focusing on concepts such as sense of place and using literature (e.g. Pocock, 1980) and landscape (Lowenthal and Prince, 1964, 1965; Appleton, 1975) as the source of 'data'. Understanding was to be achieved via the minds and actions of the principal actors in particular events, and explanation was shunned because it suggested predictability and social control rather than human freedom. The philosophies of existentialism, idealism and phenomenology were explored, but few researchers were involved and their attempts to redirect geography have been largely unsuccessful. Nevertheless, there has been a substantial volume of work and achievement in historical geography.

A second reaction to positivist human geography has its roots in the economic and social problems of British society and the apparent failure to solve them scientifically. Smith (1977), for example, argued for what he called welfare geography — 'the well-being of society as a spatially variable condition should be *the* focal point of geographical enquiry . . . the quality of lives is of paramount interest' (pp. 362-3). This largely involved mapping variables of social and economic interest, interpreting the results, accounting for their genesis, and suggesting reforms ('spatial engineering') that would lead to better lives for all.

Such a reconstructed positivist approach was criticised for the poverty of its accounts, and the inability of its proposed reforms to do more than counter the symptoms. The cause of the problems was much deeper, as argued by Harvey (1973), who contended — in his influential *Social Justice and the City* — that only Marxist theory provided both a full account of and a viable programme for change. The inequalities mapped and analysed by geographers are necessary products of a capitalist mode of production; to remove them, capitalism must be replaced by socialism.

Work in the Marxist tradition argues that human geography, unlike physical geography, cannot find universal laws of spatial behaviour, because the economic processes conditioning that behaviour are always changing — and can be changed by deliberate action. The adoption of a natural science perspective ascribes a false universality to findings — whether of a quantitative nature or not — and their use in planning and other policy programmes merely reproduces the conditions of their creation (Sayer, 1979).

Marxist research seeks to identify the economic processes that govern human activity, and to illustrate them in particular instances. It does not claim that a cause stimulates a particular effect, as positivist science does; rather, it argues that economic processes enable people to act in particular ways but constrain them from acting in others — which way is chosen is unpredictable from the general theory, and can only be understood in retrospective empirical investigations. Thus man is not presented as an automaton, but an actor operating on a pre-defined stage. Analysis of his acts has covered a range of topics in urban and economic geography (for a bibliography, see King in Quaini, 1982), and increasingly involves an integration of humanistic and Marxist perspectives (Eyles, 1981). The goal of such work, outlined in some detail by Gregory (1978), is emancipation; people are released from their condition of dependency by being made aware of their social and economic situation and determining to change that.

During the 1960s the quantitative dominance and the research emphasis on morphological regularities gave the impression of a relatively unified discipline. This was not true in the 1970s and 1980s, as researchers became more concerned with process than with form. The points of contact between physical and human geographers became of diminished importance, whereas within human geography the positivist philosophy came under severe attack — and defended itself vigorously (Hay, 1979) — while maintaining a substantial output of empirical research. Other philosophies have been proposed, including a reconstructed positivist (or welfare) approach, a humanistic perspective and a Marxist (or structuralist) epistemology (Johnston, 1983b, 1983c). All seek the same goal — an improvement in the human condition — but in very different ways.

In Search of a Focus

Geographers, like other academics, are constantly concerned about the viability and status of their discipline. To some, pluralism in approaches leads to vitality, but to others security is needed, which can be provided by identifying a disciplinary core. For many British geographers, this core must involve both physical and human geography, and the subject's identity rests on its integrating syntheses.

Prior to the period discussed here, the disciplinary core was the study of the region (Johnston, 1983a). By the 1960s this had been ousted, for a majority of geographers, by a unity based on methodology, particularly the technicalities of studying spatial patterns; the map, and its statistical analysis, was all that held the discipline together, for most systematic specialists gained greater research stimulus from the publications of other disciplines than from those of other geographers. Later, the systems approach offered a technical means of integrating studies of physical and human environments (though this was criticised as instrumentalist, treating man as a machine: Gregory, 1980).

There is now much less common ground in technical matters, as interest in processes has propelled physical and human geographers in opposite directions. To some, this separation is belated recognition of the independent entities of physical and human geography, the one closely linked to the physical and life sciences and the other to the social sciences. But among others, for both academic and political reasons, this split is deplored. Geography's special role bridging the two types of work must be maintained – to some by a revival of the regional approach, especially in teaching (Steel, 1982). And a new focus for this bridging has been identified in the study of environmental issues, widely termed resource analysis and management. Geographers, by the breadth of their training, it was argued, are uniquely placed to study man-environment interrelations, and to develop accordingly as an applied discipline (Gregory, 1974; Clayton, 1976; Trudgill, 1977b). This orientation has been promoted more by physical than by human geographers, but there has been very little integration across the bridge (Johnston, 1982). Thus physical geographers have investigated the impact of man in landscape change (Goudie, 1980) and have outlined how resource managers need to incorporate physical geography in their investigations (Cooke and Doornkamp, 1974). Human geographers have focused on the environment as a scarce resource (Parker and Penning-Rowsell, 1981), analysing how societies and individuals react to environmental constraints and resolve environmental conflicts (O'Riordan, 1981). Neither studies the processes that are the predominant research concern of the other.

One of the scarce resources of increasing interest, to society as a whole as well as geographers, is the landscape. Traditional regional geographers associated each of their homogeneous units with a characteristic landscape. Use of this concept declined in the 1950s and 1960s – although hybrid terms such as 'economic landscape' were invented and there was some investigation of townscapes – but it has recently been revived. Measurement of landscape quality, as an objective exercise, has interested physical geographers (e.g. Linton, 1968) whereas human geographers have focused on subjective assessments by observers (Penning-Rowsell, 1981), including work on landscape as an element in human self-creation and self-reflection (Appleton, 1975; Lowenthal, 1975). Both have been involved in conservation and preservation activities. Again, the laboratory is common, but the aims and procedures rather different; there is little integration of physical and human geographical research.

Applied geography

In much recent work by all geographers there is a strong utilitarian orientation, reflecting both individual and group concern with pressing world and local problems (e.g. Bennett, 1981) and a desire to preserve the discipline's strength in a period of considerable public stress on the utility of academic work. Such an applied focus is far from new, of course (House, 1973; Hall, 1981), as illustrated by Stamp's Land Utilisation Survey of the 1930s, Coleman's Second Land-Use Survey in the 1960s, and the major contribution of geographers to land-use policy and debate (Best, 1982): Stamp's introductory volume on *Applied Geography* was published in 1960, and was followed by a volume on medical geography (Stamp, 1964).

In the 1950s and 1960s geographers were numerous among the recruits to the new profession and discipline of town and country planning. Initially their focus was on land use, but later in the 1960s and through the 1970s geographical expertise has contributed substantially to the development of sophisticated models of land-use-transport interrelationships (e.g. Batty, 1976). Geographers have also been involved in the evaluation of planning policies, such as that of urban containment (Hall *et al.*, 1973), and in understanding how planning decisions are made (Blowers, 1980; Hall, 1980). To some this involvement was insufficient, and there was a feeling that geographical expertise was underrated in the corridors of power (Coppock, 1974; Steel, 1974). More applied geography was called for (and undoubtedly produced; Briggs, 1981), and a new journal, *Applied Geography*, was launched in 1981. To some this was undesirable, either because it represented a utilitarian invasion that could undermine scholarship, and the concepts of the university and academic freedom, or because it implied support for the current societal power structure and was more relevant to certain interests than to others (Harvey, 1974).

The British Impact

In many ways, British geography is an amalgam of local, North American and European trends, with the trans-Atlantic flow being particularly influential in the last 30 years. The regional focus had strong European roots, but the later systematic developments had much closer North American connections (in geomorphology with other disciplines rather than with geography) as did the spatial analysis and quantitative revolutions. Indeed, it is difficult to disentangle the British and American contributions, especially in human geography (Johnston, 1983c), for some American geographers received their first degrees in Britain, some British geographers have studied in North America, and there is substantial short-term trans-Atlantic migration. Publishing outlets tend to be shared too; there are more specialist research journals in Britain for geographers, and British publishing houses are more sympathetic to books other than the texts aimed at large first-year undergraduate markets which are popular with their North American counterparts.

Apart from the growing internationalisation of the geographical community, in which Britain has played a substantial part, there has been a major British influence on the development of academic geography departments through the provision of staff, at all levels, especially in the 1950s and 1960s. Not surprisingly, this has been greatest in the Commonwealth, both the settler-colonies (Australia, Canada and New Zealand – much less so South Africa, with its strong Afrikaner connection to continental Europe) and the more recently independent countries of South-East Asia, East and West Africa and the Caribbean: India, Pakistan and Bangladesh have employed few British geographers, however. In this way the contemporary British traditions and developments were transported to, even virtually imposed upon, a range of other countries. In addition, many of those countries, plus those of the Indian subcontinent and of the Middle East, have traditionally sent their graduates and junior staff to Britain to obtain higher degrees, and these have returned carrying the latest British research and teaching ideas. Such links are now declining. Expatriates are less frequently appointed and local postgraduate training is encouraged, in part as a reaction to large increases in British university fees. In the future, while the British impact on geography throughout the world will undoubtedly remain vital, its developments are more likely to be attuned to local circumstances and it will no longer be a source of academic manpower.

Summary

In terms of research, publications and graduate numbers, British geography has flourished in the last 35 to 40 years as never before (Johnston and Brack, 1983). Its substantive interests are wider, its methodologies more sharply honed, and its philosophical understanding greater than previously. The discipline has been lively, debates substantial and fruitful, and contributions to society many. And British geographers have participated in the development of the discipline, in a great variety of ways, in many parts of the world.

To some, this is sufficient. To others, there has been the continual worry, ever since the early 1950s, that this liveliness has been won at a serious cost in geographical unity. Physical and human geography are further apart than ever before, and within each the division into separate specialisms means that most geographers have more in common with, and indeed publish their research for, workers in other disciplines than they do with their fellow geographers. Disciplinary vitality has been bought at the cost of disciplinary coherence. To others, the concerns are less about substance than utility. Should geographers seek to be disinterested scholars, active technicians, or critical catalysts of social change? Can a single discipline contain all three? Such questions do not split geographers into mutually uncomprehending and suspicious camps, but they are the source of continuing concern among a large body of scholars with, unlike in the past, no clearly identifiable academic leaders stamping a particular image on the discipline by their own personalities and work. Perhaps British geography today has an agile mind, and a very active body, but no soul!

References

Appleton, J.H. (1975) *The Experience of Landscape*, John Wiley, London.

Atkinson, B.W. (1981a) *Meso-scale Atmospheric Circulations*, Academic Press, London.

-—— (ed.) (1981b) *Dynamical Meteorology – an introductory selection*, Methuen, London.

Balchin, W.G.V. (1952) 'Erosion surfaces of Exmoor and adjoining areas', *Geographical Journal, 118*, 453-76.

Barrett, E.C. (1967) *Viewing Weather from Space*, Longman, London.

-——(1974) *Climatology from Satellites*, Methuen, London.

-·—— and Hamilton, M.G. (1982) 'The use of geostationary, satellite data in environmental science', *Progress in Physical Geography, 6*, 159-214.

Barry, R.G. and Perry, A.W. (1973) *Synoptic Climatology: Methods and Applications*, Methuen, London.

Batty, M. (1976) *Urban Modelling: Algorithms, Calibrations, Predictions*, Cambridge University Press, Cambridge.

Baulig, H. (1935) 'The Changing Sea Level', *Transactions, Institute of British Geographers, 3.*

Bennett, R.J. (1979) *Spatial Time Series*, Pion, London.

—— (1981) 'Quantitative and theoretical geography in Western Europe', in R.J. Bennett (ed.) *European Progress in Spatial Analysis*, Pion, London, 1-32

—— and Chorley, R.J. (1979) *Environmental Systems*, Methuen, London.

Best, R.H. (1982) *Land Use and Living Space*, Methuen, London.

Blowers, A.T. (1980) *The Limits of Power*, Pergamon, Oxford.

Briggs, D.J. (1981) 'The principles and practice of applied geography', *Applied Geography, 1*, 1-8.

Brown, E.H. (1957) 'The physique of Wales', *Geographical Journal, 123*, 208-20.

—— and Waters, R.S. (1974) 'Geomorphology in the United Kingdom since the first world war' in E.H. Brown and R.S. Waters (eds.) *Progress in geomorphology: papers in honour of David L. Linton*, Institute of British Geographers, Special Publication No. 7, 3-9.

Brunsden, D.S., Doornkamp, J.C. and Jones, D.K.C. (1978) 'Applied geomorphology: a British view' in C. Embleton, D.S. Brunsden and D.K.C. Jones (eds.) *Geomorphology: present problems and future prospects*, Oxford University Press, Oxford, 251-62.

Brunsden, D.S. and Thornes, J.B. (1979) *Geomorphology and Time*, Methuen, London.

Buchanan, R.O. (1935) 'The pastoral industries of New Zealand: a study in economic geography', *Transactions, Institute of British Geographers, 2*.

Burton, I. (1963) 'The quantitative revolution and theoretical geography', *The Canadian Geographer, 7*, 151-62.

Cameron, I. (1980) *To the Furthest Ends of the Earth*, Macdonald, London.

Carson, M.A. and Kirby, M.J. (1972) *Hillslope Form and Process*, Pion, London.

Chandler, T.J. and Gregory, S. (eds.) (1976) *The Climate of the British Isles*, Longman, London.

Chorley, R.J. (1962) 'Geomorphology and general systems theory', *Professional Paper 500-B*, United States Geological Survey, Washington DC.

—— (1964) 'Geography and analogue theory', *Annals, Association of American Geographers, 54*, 127-37.

—— and Haggett, P. (eds.) (1965) *Frontiers in Geographical Teaching*, Methuen, London.

—— and Haggett, P. (eds.) (1967) *Models in Geography*, Methuen, London.

—— and Kennedy, B.A. (1971) *Physical Geography: A Systems Approach*, Prentice-Hall, London.

Clayton, K.M. (1976) 'Environmental sciences/studies', *Area, 8*, 98-101.

Cliff, A.D., Haggett, P., Ord, J.K., Bassett, K. and Davies, R.B. (1975) *Elements of Spatial Structure*, Cambridge University Press, Cambridge.

——, Haggett, P., Ord, J.K. and Versey, G.R. (1981) *Spatial Diffusion*, Cambridge University Press, Cambridge.

—— and Ord, J.K. (1973) *Spatial Autocorrelation*, Pion, London.

—— and Ord, J.K. (1981) *Spatial Processes*, Pion, London.

Conzen, M.R.G. (1960) 'Alnwick, Northumberland: a study in town-plan analysis', *Transactions, Institute of British Geographers, 27*.

Cole, M.M. (1961) *South Africa*, Methuen, London.

Cooke, R.U. and Doornkamp, J.C. (1974) *Geomorphology and Environmental Management*, Oxford University Press, Oxford.

Coppock, J.T. (1974) 'Geography and public policy: challenges, opportunities, and implications', *Transactions, Institute of British Geographers, 63*, 1-16.

Cotton, C.A. (1947) *Climatic Accidents in Landscape Making*, Whitcombe and Tombs, Christchurch, NZ.

Crowe, P.R. (1936) 'The rainfall regime of the Western Plains', *Geographical Review, 26*, 463-84.

Curtis, L.F., Courtney, F. and Trudgill, S.T. (1976) *Soils in the British Isles*, Longman, London.

Darby, H.C. (ed.) (1936) *An Historical Geography of England*, Cambridge University Press, Cambridge.

—— (1977) *Domesday England*, Cambridge University Press, Cambridge.

Dickinson, R.E. (1947) *City, Region and Regionalism*, Routledge and Kegan Paul, London.

East, W.G. and Moodie, A.E. (eds.) (1956) *This Changing World*, Methuen, London.

Edwards, K.C. (1964) 'The importance of biogeography, *Geography, 49*, 85-97.

Estall, R.C. and Buchanan, R.O. (1961) *Industrial Activity and Economic Geography*, Hutchinson, London.

Eyles, J. (1981) 'Why geography cannot be Marxist', *Environment and Planning A, 13*, 1371-88.

Eyre, S.R. (1963) *Vegetation and Soils: a world picture*, Edward Arnold, London.

Fisher, C.A. (1964) *Southeast Asia*, Methuen, London.

Fisher, W.B. (1950) *The Middle East*, Methuen, London.

Freeman, T.W. (1958) *Geography and Planning*, Hutchinson, London.

—— (1961) *A Hundred Years of Geography*, Duckworth, London.

—— *(1979)* 'The British school of geography', *Organon, 14*, 205-16.

—— (1980a) 'The Royal Geographical Society and the development of geography', in E.H. Brown (ed.), *Geography Yesterday and Tomorrow*, Oxford University Press, Oxford, 1-99.

- —— (1980b) *A History of Modern British Geography*, Longman, London.

Frisby, E.M. and Green, F.H.W. (1949) 'Further notes on comparative regional climatology', *Transactions, Institute of British Geographers, 15*, 143-51.

Gold, J.R. (1980) *An Introduction to Behavioural Geography*, Oxford University Press, Oxford.

Goudie, A. (1980) *The Human Impact*, Basil Blackwell, Oxford.

Green, F.H.W. (1950) 'Urban hinterlands in England and Wales: an analysis of bus services', *Geographical Journal, 116*, 64-88.

Gregory, D. (1978) *Ideology, Science, and Human Geography*, Hutchinson, London.

—— (1980) 'The ideology of control: systems theory and geography', *Tijdschrift voor Economische en Sociale Geografie, 71*, 327-42.

Gregory, K.J. and Walling, D.E. (1973) *Drainage Basin Form and Process*, John Wiley, London.

Gregory, S. (1963) *Statistical Methods and the Geographer*, Longman, London.

—— (1974) 'The geographer and natural resources research', *The South African Geographer, 4*, 371-82.

—— (1976) 'On geographical myths and statistical fables', *Transactions, Institute of British Geographers, NS1*, 385-400.

—— (1983) 'Quantitative geography: the British experience and the role of the Institute', *Transactions, Institute of British Geographers, NS8*, 80-9.

Haggett, P. (1965) *Locational Analysis in Human Geography*, Edward Arnold, London.

—— and Chorley, R.J. (1969) *Network Models in Geography*, Edward Arnold, London.

—— Cliff, A.D. and Frey, A.E. (1977) *Locational Analysis in Human Geography* (second edition), Edward Arnold, London.

Hall, P. *et al.* (1973) *The Containment of Urban England*, Allen and Unwin, London.

—— (1980) *Great Planning Disasters*, Weidenfeld and Nicolson, London.

——(1981) 'The geographer and society', *Geographical Journal, 147*, 145-52.

Harvey, D. (1969) *Explanation in Geography*, Edward Arnold, London.

—— (1973) *Social Justice and the City*, Edward Arnold, London.

——(1974) 'What kind of geography for what kind of public policy?', *Transactions, Institute of British Geographers, 63*, 18-24.

Hay, A.M. (1979) 'Positivism in human geography: response to critics', in D.T. Herbert and R.J. Johnston (eds.), *Geography and the Urban Environment, Volume 2*, John Wiley, London, 1-26.

House, J.W. (1973) 'Geographers, decision takers and policy makers', in M. Chisholm and H.B. Rodgers (eds.), *Studies in Human Geography*, Heinemann, London, 272-305.

Huggett, R.J. (1980) *Systems Analysis in Geography*, Oxford University Press, Oxford.

Johnston, R.J. (1982) 'Resource analysis, resource management, and the integration of physical and human geography', *Progress in Physical Geography, 6*, 127-46.

—— (1983a) 'Regional geography in Britain', in A. Kuklinski and J.G. Lambooy (eds.), *Regions, Boundaries and Spaces*, Mouton, Amsterdam.

—— (1983b) *Philosophy in Human Geography: An Introduction to Contemporary Approaches*, Edward Arnold, London.

—— (1983c) *Geography and Geographers* (second edition), Edward Arnold, London.

—— and Brack, E.V. (1983) 'Appointment and promotion in the academic labour market: a preliminary survey of British University Departments of Geography, 1933-1982', *Transactions, Institute of British Geographers, NS8*, 100-11.

Jones, D.K.C. (1981) *The Geomorphology of the British Isles: Southeast and Southern England*, Methuen, London.

Jones, E. (1956) 'Cause and effect in human geography', *Annals, Association of American Geographers, 46*, 369-77.

Jones, E. (1960) *A Social Geography of Belfast*, Oxford University Press, Oxford.

King, L.C. (1953) 'Canons of landscape evolution', *Bulletin of the Geological Society of America, 64*, 721-52.

Lamb, H.H. (1950) 'Types and spells of weather around the year in the British Isles: annual trends, seasonal structure of the year, singularities', *Quarterly Journal of the Royal Meteorological Society,*

76, 393-429.

Leopold, L.B. and Langbein, W.B. (1962) 'The concept of entropy in landscape evolution', *Professional Paper 500-A*, United States Geological Survey, Washington DC.

—— Wolman, M.G. and Miller, J.P. (1964) *Fluvial Processes in Geomorphology*, Freeman, San Francisco.

Linton, D.L. (1968) 'The assessment of scenery as a natural resource', *Scottish Geographical Magazine, 84*, 219-38.

Lockwood, J.G. (1974) *World Climatology: an environmental approach*, Edward Arnold, London.

Lowenthal, D. (1975) 'Past time, present place: landscape and memory', *Geographical Review, 65*, 1-36.

—— and Prince, H.C. (1964) 'The English landscape', *Geographical Review, 54*, 309-46.

—— and Prince, H.C. (1965) 'English landscape tastes', *Geographical Review, 55*, 186-222.

Martin, A.F. (1951) 'The necessity for determinism', *Transactions, Institute of British Geographers, 17*, 1-12.

Oldfield, F. (1960) 'Late quaternary changes in climate, vegetation and sea-level in lowland Lonsdale', *Transactions, Institute of British Geographers, 28*, 99-117.

O'Riordan, T. (1981) *Environmentalism*, Pion, London.

Parker, D.J. and Penning-Rowsell, E.C. (1981) *Water in Britain*, Allen and Unwin, London.

Pears, N.V. (1977) *Basic Biogeography*, Longman, London.

Peltier, L.C. (1950) 'The geographic cycle in periglacial regions as it is related to climatic geomorphology', *Annals, Association of American Geographers, 40*, 214-36.

Penman, H.L. (1948) 'Natural evaporation from open water, bare soil and grass', *Proceedings of the Royal Society, Series A, 193*, 120-45.

Penning-Rowsell, E.C. (1981) 'Fluctuating fortunes in gauging landscape value', *Progress in Human Geography, 5*, 25-41.

Pocock, D.C.D. (ed.) (1980) *Humanistic Geography and Literature*, Croom Helm, London.

Quaini, M. (1982) *Geography and Marxism*, Basil Blackwell, Oxford.

Rawstron, E.M. (1958) 'Three principles of industrial location', *Transactions, Institute of British Geographers, 25*, 135-42.

Rhind, D.W. and Adams, T.A. (1980) 'Recent developments in surveying and mapping' in E.H. Brown (ed.), *Geography Yesterday and Tomorrow*, Oxford University Press, Oxford, 181-99.

Sayer, R.A. (1979) 'Epistemology and conceptions of people and nature in geography', *Geoforum, 10*, 19-44.

Scheidigger, A.E. and Langbein, W.B. (1966) 'Probability concepts in geomorphology', *Professional Paper 500-C*, United States Geological Survey, Washington DC.

Schumm, S.A. and Lichty, R.W. (1965) 'Time, space and causality in geomorphology', *American Journal of Science, 263*, 110-19.

Smailes, A.E. (1946) 'The urban mesh of England and Wales', *Transactions, Institute of British Geographers, 11*, 85-101.

—— (1953) *The Geography of Towns*, Hutchinson, London.

Smith, D.M. (1977) *Human Geography: A Welfare Approach*, Edward Arnold, London.

Smith, W. (1952) *An Economic Geography of Great Britain*, Methuen, London.

Spate, O.H.K. (1954) *India and Pakistan*, Methuen, London.

—— (1957) 'How determined is possibilism?', *Geographical Studies, 4*, 3-12.

Stamp, L.D. (1947) *The Land of Britain*, Longman, London.

—— (1964) *Some Aspects of Medical Geography*, Oxford University Press, Oxford.

—— (1966) 'Ten years on', *Transactions, Institute of British Geographers, 40*, 11-20. ·

Steel, R.W. (1974) 'The Third World: geography in practice', *Geography, 59*, 189-207.

—— (1982) 'Regional geography in practice', *Geography, 67*, 2-8.

Stoddart, D.R. (1965) 'Geography and the ecological approach: the ecosystem as a geographical principle and method', *Geography, 50*, 242-51.

—— (1967) 'Organism and ecosystem as geographical models' in R.J. Chorley and P. Haggett (eds.), *Models in Geography*, Methuen, London, 511-48.

—— (ed.) (1981a) *Geography, Ideology and Social Concern*, Basil Blackwell, Oxford.

—— (1981b) 'Geography, education and research', *Geographical Journal, 147*, 287-97.

Strahler, A.N. (1964) 'Quantitative geomorphology of drainage basins and channel networks', in V.T. Chow (ed.) *Handbook of Applied Hydrology*, McGraw-Hill, New York, 4-76.

Taylor, P.J. (1976) 'An interpretation of the quantification debate in British geography', *Transactions, Institute of British Geographers, NS1*, 129-42.

Thomas, M.F. (1972) *Tropical Geomorphology*, Macmillan, London.

Thomas, R.W. and Huggett, R. (1980) *Modelling in Geography*, Harper and Row, London.

Tivy, J. (1971) *Biogeography, a study of plants in the ecosphere*, Oliver and Boyd, Edinburgh.

Trudgill, S.T. (1977a) 'Environmental sciences/studies: depth and breadth in the curriculum', *Area, 9*, 266-71.

—— (1977b) *Soil and Vegetation Systems*, Oxford University Press, Oxford.

Unwin, D.J. (1977) 'Statistical methods in physical geography', *Progress in Physical Geography, 1*, 185-221.

Ward, R.C. (1967) *Introduction to Hydrology*, Macmillan, London.

Waters, R.S. (1958) 'Morphological mapping', *Geography, 43*, 10-17.

Watts, D. (1971) *Principles of Biogeography*, McGraw-Hill, London.

Whitehand, J.W.R. (ed.) (1982) *The Urban Landscape, Historical Development and Management*, Academic Press, London.

Wilson, A.G. (1970) *Entropy in Urban and Regional Modelling*, Pion, London.

—— (1981a) *Catastrophe Theory and Bifurcation: Applications to Urban and Regional Systems*, Croom Helm, London.

—— (1981b) *Environmental Systems Analysis*, John Wiley, Chichester.

Wooldridge, S.W. and East, W.G. (1951) *Spirit and Purpose of Geography*, Hutchinson, London.

—— and Linton, D.L. (1939) 'Structure, surface and drainage in South-East England', *Transactions, Institute of British Geographers, 10*.

Worsley, P. (1979) 'Whither geomorphology?', *Area, 11*, 97-101.

Wrigley, N. and Bennett, R.J. (eds.) (1981) *Quantitative Geography*, Routledge and Kegan Paul, London.

Young, A. (1972) *Slopes*, Oliver and Boyd, Edinburgh.

7 POLAND

Antoni Kuklinski

This is not a conventional paper on the history of Polish geography (for that, see Kuklinski, 1983b). Rather, it is an attempt to present to the international geographical community the beginnings of new approaches now emerging in Poland and of analyses and evaluations of the experience and prospects of Polish geography. These new approaches are well documented in a special issue of the *Polish Geographical Review*, to be published at the end of 1983 (in Polish, with Russian and English summaries).[1]

Such a paper should not be considered as presenting the conventional wisdom regarding the history of Polish geography since the Second World War. Many of the points and judgements included have a clearly subjective and sometimes controversial character. Thus the critical reader should test the content of the paper against the rich and excellent collection of material published since 1964 in *Geographica Polonica* (the journal of the Polish Academy of Sciences). Fifty volumes have been published, mainly in English but with a number of issues in French. The present paper is perhaps best described as a nonconventional footnote to those 50 volumes.

Two Models

Studies of the history of Polish geography since the Second World War have been dominated by three types of classical descriptive analysis: (1) those that describe research activity in the discipline as a whole, and in its various subdisciplines (such as geomorphology and settlement geography); (2) those that describe the activities of the various institutions, such as the Institute of Geography of the Polish Academy of Sciences and the Polish Geographical Society; and (3) those that describe the work of eminent geographers. In these descriptions one can identify, though not without some simplification, a particular concept of development. This is defined as a process of permanent accumulation of academic achievements as an uninterrupted sequence of successes and positive changes. In addition, it is seen as a harmonious process, thereby disregarding not only interpersonal disagreements and antagonisms but also major academic debates that are among the most important driving forces in the progress of any academic discipline.

Three questions can be formulated which test the validity of the model implied in the classical descriptive approaches:

(1)   Is it possible to describe the development of Polish geomorphology without reference to the very vivid polemic between Jan Dylik and Mieczyslaw Klimaszewski, a polemic that was one of the most interesting aspects of that subdiscipline in the 1950s?

(2)   Is it possible to describe developments in social and economic geography without mentioning the almost complete breakdown in these fields in 1950-4?

(3)   Is it possible to describe the academic activity of eminent Polish geographers indicating their positive achievements only and mentioning not even one example of a paper including a factographic error, a false hypothesis or a misguided prognosis?

All three must be answered, but in the classical model such questions have not even been formulated.

I propose to abandon the classical model and to replace it by a paradigmatic approach (c.f. Kuklinski, 1983d) which will generate new questions to be addressed to the history of Polish geography. The model to be developed is inspired by the work of Kuhn (1962), although this provides a general perspective only and not a rigorous analytical framework. Its fundamental features include:

(1)   Development is presented as a cyclical phenomenon, incorporating periods of stagnation, even recession, as well as of progress.

(2)   Development is polarised, so that within geography different schools emerge, blossom and decline, producing a strong differentiation of innovation with subdisciplines that are either intellectual growth poles, areas of average dynamics, or dying.

(3)   In the development process a special role is given to academic debate and to the emergence and resolution of conflicts relating to methodological and empirical progress – a resolution that requires a mechanism.

(4)   Development is both a *laissez-faire* and a guided phenomenon.[2]

(5)   In studying the development process, special attention must be paid to the forces and conditions that generate fundamental changes in the academic and social status of the discipline.

In putting such a model into operation, it is necessary to ask whether scientific revolutions have taken place in Polish geography since the Second World War. If they have, then the history of those revolutions must be analysed in the context of international academic stimuli and of general political, social and economic changes within Poland.

This paradigm model is a rudimentary one, but it contains the seeds of innovative thinking about the recent experience of Polish geography. This thinking can be stimulated by two recent intellectual streams in international academic writing: a reflection on the styles and mechanisms of socio-economic development (e.g. Wolfe, 1981); and reflections on fundamental problems in the development of natural, social and humanistic sciences (e.g. Ziman, 1968; Johnston, 1979; Komorowski, 1982; Chojnicki, 1983). Here, I will concentrate on an analysis of the mechanism of development in Polish geography, an analysis which should bridge the classical and paradigm models.

Mechanisms of Development

No attempt will be made here to define the notion of a mechanism of development. Rather, I will outline in an intuitive way and then analyse four mechanisms which are useful for understanding developments in Polish geography.

Creative Thinking

Creative thinking is the most important mechanism of development in all academic disciplines. Two approaches could be taken to its analysis, in both methodological and empirical studies (Mikulinski and Jaroszewski, 1969; Cackowski, 1972): the classical approach, concentrating on the results of research activity, notably in the form of publications; and recent psychological, sociological and historical approaches which concentrate on the personalities of scholars, on the conditions that shape their creative abilities, and on the social mechanisms that select creative scholars in particular historical situations. The former could be called the *output approach* and the latter the *input approach*.

Analysis of the materials in the special issue of the *Polish Geographical Review* already referred to leads one to conclude that four outstanding examples of creative thinking in Polish geography since 1945 are worthy of mention. They refer to periglacial geomorphology, to geomorphological mapping (Gilewska, Klimkowa and Starkel, 1982), to the Polish settlement system and to agricultural land use and typologies (Jankowski, 1982). A comprehensive analysis of none is presented here, but four important features are listed as indicators of the common elements in such creative thinking. The first feature is the fundamental role played by eminent scholars in the innovatory work in each. The second is the ability to organise and inspire academic teams to implement and develop new fields and place them on proper empirical foundations. Third is the positive attitude towards such work by the state, which provided the necessary financial and institutional conditions. Finally, there is the positive role played by international cooperation, which incorporated Polish work into the mainstreams of geographical thinking; the stimulating environment of the International Geographical Union deserves special mention.

These four examples do not comprise a comprehensive list of all creative thinking in Polish geography. I would suggest that a full list would contain about 20 cases, but further study is needed to evaluate that claim.

There are two other issues related to the mechanism of creative thinking. The first concerns an evaluation of the national system of geographical education, from kindergarten to PhD, and asks whether that system is stimulating or repressing creative geographical thinking. For this evaluation we need historical, international and cross-disciplinary comparisons. My fear is that, for Poland, a cross-disciplinary comparison with, say, physics would not produce favourable results with respect to geographical education.

The next issue is fundamental to both *laissez-faire* and guided approaches to development in Polish geography; how do the mechanisms of selection, promotion and obliteration operate with respect to people who demonstrate exceptional abilities in creative, non-conventional thinking? Are such mechanisms infiltrated by strong strata of mediocrity and conformism, creating effective barriers to the emergence and development of eminent academic personalities? Do the mechanisms provide avenues along which young scholars could steer new vehicles of creative and innovative thinking?

It is doubtful whether studies could be formulated to answer these questions – in any discipline. Nevertheless, the questions should be asked, in the expectation that indirect answers can indicate the internal processes that shape our minds, hearts and consciences. For Polish geography, I do not suggest that such indirect answers will be either more pessimistic or less optimistic than is the case in other countries; since the Second World War there have been many examples of creative thinking that have been incorporated into the mainstream of international disciplinary progress.

Information Creation and Absorption

The mechanism for the creation and absorption of information is an important driving force in the development of geography. There are two relevant issues: the creation of information by the discipline itself; and the absorption and transformation of information generated by institutions external to the discipline.

There are three major examples of the *creation of new sources of information within Polish geography* since the Second World War. These are investigated here from the viewpoint of the theory of information. It is suggested that in this context both the geomorphological and the land-use map have been successes for Polish geography, because of two factors, but the hydrographic map has not been a scientific success. With the first two map types, the compilers have accumulated a sufficient foundation of theoretical concepts and research hypotheses so that valid questions could be answered by the maps. In addition, the observational and measurement instruments used to prepare the maps were more or less in conformity with elementary international standards. Neither factor was available to promote the success of the hydrographic map. Thus geomorphological and land-use mapping provided informational foundations for the reconstruction of Polish geomorphology and agricultural geography respectively. The limited success of the hydrographical mapping was one of the reasons for the weak development of Polish hydrology as a geographical discipline in the 1950s and 1960s; more recently, mapping has been considered relatively unimportant as a means of methodological reconstruction in this area.

These examples of past developments built on new informational sources can be complemented by prognoses concerning the likely impact of new sources in the 1980s. I believe that there will be two new intellectual growth poles in the next decade — in social and in cultural geography. It is difficult to develop imaginative programmes in these fields without the creation of new sources of information adapted to the needs of geographical research. These should include the application of different types of questionnaire, reducing geographers' one-sided reliance on conventional statistical data and creating independent data sources for their research.

Many geographers in Poland have already used questionnaires in their research, but too little attention has been given to the general evaluation of these tools and their role in information creation. The relative weakness of geography in this field can be contrasted with the vast sociological literature. But it would be a mistake to suggest an automatic transfer of sociological experience of questionnaires to geography; we have to develop our own empirical and methodological experience and expertise.

Just as the creation of endogenous information sources plays a major role in the analysis of objective reality and the methodological reconstruction of Polish geography, so too does *the absorption and transformation of information created by external institutions*. In the history of the discipline there are many examples of this process, with both negative and positive impacts. The performance of such information can be assessed in terms of the quantity, quality and accessibility of cartographical, statistical and bibliographical sources.

The technological and institutional restrictions on the accessibility of cartographic and remote-sensing data have had a negative impact on Polish geography, limiting developments in an international context. The quantity, quality and accessibility of statistical data have not had such an unequivocal impact (except in the period 1950-65, when the Central Statistical Office was on the verge of extinction); overall the Polish statistics are reasonable, in scope and intensity, compared to those in other countries. There are even fields — investment statistics, for example — in which Polish data are superior to those of Western countries, where strict rules of competition apply and secrecy is the norm. In the 1960s and 1970s, therefore, Polish regional statistics were comparatively rich. Today, collection of those statistics is experiencing a deep methodological, technological and managerial crisis, and Polish geographers are participating in discussions related to the recovery and reconstruction of Polish regional statistical sources.

During the years 1956-81 accessibility to the vast range of international publications and journals was not a problem for Polish geography. Libraries, particularly the Central Library in Warsaw, were well-founded, on an international scale of comparison. At present, currency restrictions imposed because of the economic crisis have reduced the flow of materials, but it is to be hoped that the *status quo ante* in the supply of foreign books and journals will soon be re-established.

The mechanisms for creating and absorbing information are crucial to disciplinary progress. The incorporation of elements of information theory to the methodological framework of geography could provide an effective integration channel, creating common denominators among the differentiated and dispersed fields that form a loose coalition within the discipline.

Co-operation in Practical Activities

Co-operation in practical activities is an important developmental mechanism for geography in all countries (Dziewonski, 1974; Jedraszko, 1983). The positive influence of such co-operation produces: (1) intellectual and moral satisfaction for the scholar involved in the transformation of objective reality according to the accepted system of values; (2) stimulation for more effective development of research activities, including the generation of problem-orientated approaches, the establishment of interdisciplinary links, and an acceleration of the rhythm of research activity; and (3) stronger financial foundations for academic work. Very often, however, co-operation has left a bitter taste of experience with all parties, especially with regard to intellectual and moral satisfaction. Against this feeling among academics, however, must be set the complaint of practical institutions that the academic machinery moves very slowly and that the Hamletic behaviour of many scholars can be sometimes disappointing, even irritating.

In this context, it is valuable to assess the intentions and motivations of both geographers and practitioners. These two groups do not always interpret the main goals and assumptions of a co-operative venture in the same spirit. Figure 7.1 suggests four ideal-type situations, all of which are represented in the recent experience of Polish geography. Unfortunately, the first type has not dominated and was mainly experienced in periods of general enthusiasm, as in the years 1945-8 and 1950-60. In the reconstruction period (Leszcsycki, 1972) and in the period of Polish October, enthusiasm has stimulated strong intentions and motivations for real co-operation. In other times, although type 1 has not been entirely absent, the other three situations emerged relatively frequently. (In the second type, both parties perform a ceremony of fictitious co-operation.)

The third type is very common, not only in Poland, and it is not difficult to find examples of the following type of behaviour. A good, competent scholar is interested in solving a complex problem in fundamental research, but cannot get proper financial support for himself and his research team. So he surrenders to the necessity of the technological procedure by dressing up the fundamental problem in the clothes of applied research in order to incorporate it with the mechanism of practical co-operation. The practitioners react positively. The result may not be negative, for the research results may make a substantial contribution to geographical progress and the dynamic of the co-operation may shift the research from type 3 into type 1.

Contrary to the above description, it is rare to find examples of an

Figure 7.1: Intentions and Motivations of Co-operation Between the Sphere of Geography and the Sphere of Practice

		SPHERE of PRACTICE Intentions and Motivations	
		of real co-operation	of fictitious co-operation
SPHERE of GEOGRAPHY Intentions and Motivations	of real co-operation	1	4
	of fictitious co-operation	3	2

intellectual and moral recovery in the fourth type. The classic example is of an expert evaluation organised by the sphere of practice after the real closure of the decision-making process. If the decision is correct, then the only loss is the financial one of employing the expert (the geographer), who will not influence the decision. But if the decision is a wrong one, then the expert evaluation creates intellectual and moral problems for both parties. If the scholar-expert continues to be motivated towards co-operation, then the likely reaction is disappointment and frustration, but if he understands that his expert evaluation is fictitious, and can accept this, the situation moves from type 4 to type 2.

This brief outline of the intentions and motivations of the spheres of geography and practice should provide a basis for further, more comprehensive studies of the history of Polish geography. Co-operation is important to disciplinary development. Its analysis should focus not only on conventional declarations and carefully-chosen positive experiences but also on the negative experiences on both sides.

International Co-operation

No academic discipline, especially one in a small country, can be developed in isolation from the international academic community. In Poland, however, international geographical contacts were very weak during the decade 1945-55; indeed they were on the borders of isolation and extinction. The important breakthrough occurred in October 1956 when, for the first time since the Second World War, an official Polish delegation participated in the International Geographical Congress.

The Polish October set the mechanism of international co-operation in motion. Its importance to disciplinary development occurs on both the import and the export side of the balance sheet. Two types of transaction are important: the import-export of creative thinking (hypotheses, methodologies, instruments, etc.); and the import-export of factual knowledge. If a discipline is actively participating in the international community then exports should be preponderant in this trade; if it is a passive member, imports will dominate.

With regard to Polish geography in the period 1956-83 it is necessary to understand in which fields it was an active participant and in which it was a passive recipient only. Allied to this, it is necessary to study both the diffusion of innovations through Polish geography and the diffusion of Polish exports to other countries. At present only an intuitive sketch can be presented, as a proposal for appropriate studies; the judgement is that the Polish School of Geography has been a positive participant, although anxieties have been raised recently regarding an apparent decline of Polish geographical influence (Kostrowicki, 1981), as in the workings of the International Geographical Union (IGU) and its various Commissions and Working Parties.

International co-operation occurs in both multilateral and bilateral contexts. The IGU provides the best example of the former, but it is not the only one; the UN and its many institutions provide funds for international co-operation in many fields, in which Polish geographers have participated. At the bilateral level, international geographical seminars have been the major activities, with about 50 taking place in the years 1959-83. These can be evaluated in terms of a football match; which team performed best in the successive competitions? For the seven British-Polish seminars, those in the 1970s showed that the spirit of innovative expansion was much stronger in British geography.

The Developmental Ideology

The Programmes of the 1950s

Three elements of an ideology of development emerged in Polish geography during the 1950s. The first was the idea of guided development, the second that of continuing development and the third that of comprehensive development. The *idea of guided development* functioned relatively well, integrating the advantages of central planning with those of the innovative power and elasticity obtained from decentralised, autonomous, spontaneous activities. The good performance of this mechanism provided one of the necessary conditions for the annihilation of paradigm I in Polish geography and its replacement, responding to the new situation at home and abroad, by paradigm II. But its efficiency declined in the late 1960s, so that in the 1970s the spontaneous developments were much more important than those that were centrally guided.

During this period of guided development, the main navigator was Stanislav Leszczycki. The efficiency of the central mechanism reflects strongly on his personal managerial and political power and weakness. Today, I believe that such a mechanism should be reconstructed, as a necessary condition for a full renaissance of Polish geography in the 1980s, but this controversial point of view is unlikely to win majority support.

During the 1950s, in most countries, there was a dominant attitude presenting the future in the perspective of a process of *continuing development*. Polish geography was interpreted in this light, as a continuous sequence of positive qualitative and quantitative changes generating permanent improvements in disciplinary status. But the years 1960-83 have demonstrated that cyclical trends, incorporating periods of prosperity and depression, have not been eliminated from the history of society, economy, science and culture. Thus the modern history of Polish geography must be interpreted from the perspective of cyclical development too.

In the present Polish situation, this is an optimistic attitude, embracing the view that after the current depression a period of prosperity will emerge and will be reflected in Polish geography. This will not follow automatically, however; its achievement demands an efficient, well-organised effort by Polish geographers.

A third attitude promoted in Polish geography during the 1950s

Table 7.1: The Polarised Development of Polish Geography, 1960-80

GEOGRAPHICAL DISCIPLINES		I Intellectual growth poles	II Areas of average dynamics	III Areas of declining dynamics	IV Embryonic growth poles	V Pre-embryonic growth poles
PHYSICAL GEOGRAPHY	1 geomorphology	●				
	2 hydrography		●			
	3 climatology			●		
	4 geography of soils					●
	5 biogeography				●	
ECONOMIC GEOGRAPHY	6 geography of industry	●				
	7 geography of agriculture		●			
	8 geography of transport		●			
	9 geography of population		●			
	10 geography of settlement	●				
	11 historical geography			●		
	12 regional geography			●		
	13 cartography			●		

was that of *comprehensive development*, implying that all parts of the discipline should grow at comparable rates. Naturally, nobody promoted the idea of a mechanical equalisation along the frontiers of research progress, but it was accepted that resources for development should not be distributed in a disparate fashion and that there should be no underdeveloped and declining areas within the disciplinary space. Nevertheless, reality during the 1960s and 1970s conformed much more to a model of polarised development, around intellectual growth poles, than to one of comprehensive development.

The Reality of the 1960s and 1970s

In a retrospective analysis, the high degree of polarisation of development during the years 1960-83 can be illustrated with reference to five areas within the discipline (Table 7.1), differentiated according to the scale and dynamic of development. The first area is dominated by the three major growth poles of Polish geography – geomorphology, geography of agriculture and geography of settlement. Their development displays all of the characteristics of growth poles (Kuklinski, 1972); the dynamics of innovation, multiplier effects, spread and backwash effects, and internal and external linkage mechanisms.

The second area comprises four subdisciplines which display an average level of developmental dynamism – hydrography, geography of industry, geography of transport and geography of population. The third area – containing climatology, historical geography, regional geography and cartography – is one of decline, whereas the fourth contains only biogeography, an embryonic growth pole which unfortunately has had an unchanged status throughout the period. Finally there is the pre-embryonic area represented by the geography of soils; there were strongly expressed intentions to develop this in the 1950s, but practically nothing has been done to implement those intentions.

Table 7.2 presents the essential elements of this classification. Not all elements of the discipline of geography are incorporated there, because the aim has not been a full enumeration of the characteristics of Polish geography but rather the creation of a sufficient empirical base for an analysis of the process of polarised development. The location of each subdiscipline shown in the table to a developmental category reflects value judgements which are biased by subjective and insufficient knowledge about certain segments of Polish geography. The categorisation may be challenged, and should perhaps be taken as a hypothesis that can be tested against the 50 volumes of *Geographia Polonica*. Nevertheless, whatever the details of the categorisation, I would strongly defend the validity of the model of polarised development, and present it as a major instrument for the interpretation of the historical and prognostic experiences of Polish geography.

The First Scientific Revolution, the 1950s

Twice in the history of Polish geography since the Second World War a combination of factors has been created that could justifiably be termed a scientific revolution. The first of these occurred in the 1950s.

Facts, Phenomena and Processes

The main elements of this first revolution unfold as follows. Firstly, the papers presented at the Congress of the Polish Geographical Society in Gdansk in 1949 started the process of annihilation of the paradigm I of Reborn Poland, that provided the discipline's methodological foundation for the years 1918-48. Two years later, at the first Congress of Polish Science, a critical evaluation of the development of Polish scientific work began, in the context of a new programme aligning science to the new model of state, society and economy adopted in Poland after the Second World War. It was in this context that the annihilation of paradigm I in Polish geography was completed, especially for non-physical and non-cartographic fields. At the same time, the processes that created paradigm II in Polish physical geography (especially geomorphology) were initiated.

The Institute of Geography of the Polish Academy of Sciences was created in 1953 and immediately assumed the role as the main national research centre. It was granted a set of monopolistic privileges and with these could create bold development strategies. It absorbed innovative processes, created others and became the main driving force of Polish geography, leading to the emergence of paradigm II. This emergence was heralded in physical geography by the methodological and empirical debates of 1952-5 and a sequence of excellent geomorphological conferences. In economic geography, a similar reconstruction of research programmes and university curricula followed a methodological conference held at Osieczna in 1955. During this period, too, international co-operation was being advanced, via the participation of the Polish delegation at the 1956 International Geographical Congress in Rio de Janeiro and the first bilateral seminar (Polish-British) held in 1959.

Finally, between 1955 and 1960 co-operation between the sphere of geography and the sphere of practice in the field of perspective planning provided an atmosphere in which innovation flourished. The global dimension of the model of perspective planning was developed by Michal Kalecki, Kazimierz Secomski and Witold Lissowski; the regional dimension was provided by Kazimierz Dziewonski and Jozef Zaremba. In 1958, the Polish Academy of Sciences created a Committee for Space Economy and Regional Planning; this provided an effective framework for co-operation among economic geographers, planning institutions and other disciplines involved in regional studies (Kuklinski, 1966).

This selection of facts, phenomena and processes associated with the course of the first scientific revolution in Polish geography may be a biased one. It should, however, provide an inducement for a more comprehensive historical analysis – an analysis which, I am convinced, will not question the fundamental argument that there was a scientific revolution in Polish geography between 1949 and 1960, which created paradigm II in the discipline's history.

Driving Forces of the Revolution

The following were the main forces driving the revolution of the 1950s (Kuklinski, 1983d). Firstly, there was the atmosphere of creative and disciplined enthusiasm that characterised the epoch of the Polish October. Secondly, there was the substantial opening of international contacts. Thirdly, adoption of the principle of specialisation in geographical research significantly raised the qualifications and competence of Polish geographers, strengthening their ability to compete at home and abroad. Fourthly, the efficient mechanism linking theory to practice protected the discipline against both the trap of an ivory tower and that of servile eulogies. Finally, there was effective intergenerational co-operation, integrating the wisdom, tolerance and experience of the older generation with the ardour, energy and enthusiasm of younger geographers.

Evaluation

The scientific revolution of the 1950s was a success because it was able to solve three problems. First, it established the priority of *substantial elements over instrumental elements* in the construction of paradigm II – it was assumed, correctly, that formulation of a new paradigm requires a new set of questions addressed to objective reality. In the 1950s that set of questions provided the framework for a new research programme in Polish geography, putting in motion a strong set of innovative stimuli relating to creative thinking, interdisciplinary and international co-operation, and practical co-operation. By contrast, changes in the discipline's instrumental and methodological equipment were smaller, though sufficient to ensure the revolution's success.

The second problem solved was the *integration of the annihilation of paradigm I with the creation of paradigm II*. An important element of the revolution was its positive qualities. Even in the most difficult years of the early 1950s, the leadership of the revolution never concentrated much attention on the negative aspects of paradigm I. Annihilation of that paradigm was also accompanied by parallel creative thinking about the new model. Such parallelism of negative and positive work was fully implemented in physical geography. In economic geography (for reasons not dependent on the Polish geographical establishment's good intentions) there was a 'development gap' during 1950-4, but this was eliminated in the ensuing very active period.

Much of this successful solution of the problems of integrating old and new, negative and positive, appears to stem from the Osieczna Conference. The programme of positive activities formulated there had a high implementation potential because the old paradigm had already been annihilated and could generate no resistance against the forces of change.

'Difficile est mutare in melius.' The wisdom of this maxim is reflected in the solution to the final problem, *the integration of spontaneous activities and the mechanisms of guided change*. In the 1950s there was spontaneous enthusiasm among both young and old Polish geographers (though in the context of 1983 it should be noted that the old of the 1950s were very young!) That enthusiasm was not anarchic, however. It was guided by proper personal, institutional and material instruments, and the efficiency of the guiding was a fundamental condition for the success of the first scientific revolution.

Paradigm II Flourishes and Declines, 1961-80

The 20-year period 1961-80 saw paradigm II flourish and then decline. The early 1960s provided a long series of successes for Polish geography, at home and abroad, climaxed by the election of Stanislaw Leszczycki as President of the IGU at the New Delhi Congress in 1968.

During the mid-1960s the first signs of an attenuation of the driving forces of paradigm II appeared. There were two failures to realise potentials for further advance. The first concerned the mathematisation of geography and the introduction of the 'quantitative revolution' from other countries. The other related to the integration of the various geographical specialisms, for specialisation had encouraged interdisciplinary ties at the cost of internal cohesion. But neither potential was capitalised upon sufficiently to inhibit the forces that were generating a decline in paradigm II.

The 1970s brought two new sources of inspiration to Polish geography. The first was methodological, and consisted of attempts to strengthen the research apparatus by the introduction of the conceptual and methodological frameworks of systems analysis. The ideals were not realised, however; in practice application of systems analysis proved difficult and relatively few studies were produced. As a reaction, it seemed that systems analysis provided a flattering of readers with a new ornamental language rather than a more profound understanding of objective reality.

The second inspiration was institutional, for some geographers hoped that new arrangements in the finance and organisation of scientific research, in the context of centrally-managed problems, could reinvigorate geographical research. Polish geographers were indeed active in this area during the 1970s but the significant results that they obtained neither inhibited the process of decline of paradigm II nor provided any impulse for a new paradigm III.

The stars of Polish geography lost something of their brilliance in the international firmament during this decade. This does not mean that Polish geography regressed, but it does indicate (as a comparison of British and Polish geography in the 1970s makes clear) that developments were proceeding more rapidly elsewhere. In Poland by the mid-1970s the phenomena of accelerated decline could be seen (at least by a growing minority of geographers) and the elements of paradigm III begin to emerge; the second scientific revolution in Polish geography was being set in motion.

The Second Scientific Revolution

It is very difficult at present for a person deeply involved in the relevant discussions and debates relating to the controversial evaluations of the experience and prospects of geography and regional studies (e.g. Kuklinski, 1983a, 1983b, 1983c, 1983d, 1984) to analyse the progress of the second scientific revolution; I hope that I might be invited to in the late 1980s. Nevertheless, at this time a subjective statement is preferable to silence, and is presented as a hypothesis to be tested at the end of the revolution. This personal statement covers two main issues.

The Dialectics of Questions and Answers

The development of an academic discipline involves a process of permanent confrontation between discipline and problems emerging in objective reality. The nature of the confrontation can be analysed either in terms of questions generated by objective reality or in terms of those addressed to it by the discipline; in both cases the answers come from disciplinary research activity.

It is necessary in this context to distinguish between old and new questions and old and new answers. Four typical situations are outlined in Figure 7.2. I would argue that when a paradigm is dying types 1 and 2 are the most important, whereas in the framework of a growing paradigm types 3 and 4 dominate. In Poland at present, the competitive coexistence of paradigms II and III means that all four are represented. Focus here is on type 4, however, on the new questions being asked by Polish geographers. They include:

(1)   How can the ecological crisis be explained and how can a prognosis of the long-term consequences of that crisis for Polish economy and society be prepared? (Some alarmed scholars now use the term 'ecological catastrophe' when referring to some areas of Poland: Kuklinski, 1984.)

(2)   How, in a critical perspective, can the fundamental features of the industrialisation model implemented in Poland since the Second World War be explained, and how can the quality of that model be related to the quality of the Polish spatial structure in the early 1980s?

(3)   How can the fundamental features, including spatial differentiation, in the processes of social crisis and reconstruction in Poland be explained?

(4)   How can the parallel fundamental features in the country's economic crisis and reconstruction be explained?

(5)   How are new local and regional communities in Poland developing?

(6)   How can the new phenomena in the changing system of values be explained and related to their impact on the country's cultural

Figure 7.2: The Interaction of Questions and Answers in the Development of Polish Geography in the Early 1980s

		ANSWERS	
		old	new
QUESTIONS	old	1	3
	new	2	4
Paradigm		II	III

geography?

(7) How can the fundamental problems of the territorial organisation of the state be explained?

This list could be expanded in a variety of ways, depending upon the interests, professional abilities and ideological attitudes of their formulators.

There is no doubt that the dialectic of new questions and answers is the main characteristic of Polish geography in the 1980s. This dialectic is the main driving force of the second scientific revolution and the construction of paradigm III.

New Cross-Disciplinary Inspiration

Fifty years ago physics was widely recognised as the ideal methodological model for all disciplines. Today the frontiers of human knowledge are seen more vividly in biology, so that the theoretical, methodological and empirical achievements of biology provide a major source of cross-

Table 7.2: The Polarised Development of Polish Geography in the 1980s

Potential growth poles	New questions addressed to the objective reality	New cross-disciplinary inspirations for geography
Physical geography Biological geography	The origin and long-term consequences of the ecological crisis of Polish space in the perspective of international comparative studies.	The methodological and empirical achievements of ecological, biological and hydrological sciences.
Social geography Cultural geography	The geography of the social crisis and social reconstruction in Poland in the perspective of international comparative studies. The new social and cultural geography of Poland in the perspective of changing system of values, changing quality of life and new trends in the development of local and regional communities.	The theory of social and cultural transformations, sociology, psychology, cultural anthropology and social ecology.
Economic geography	The geography of economic crisis and economic reconstruction in the perspective of international comparative studies. The new geography of prices. The new geography of the enterprise.	The political economy of economic crisis and economic reconstruction. The economics of shortages. The managerial sciences.
Political geography	The territorial organisation of the State. The experience of Poland in the perspective of international comparative studies. The territorial self-government.	Politology. The theory of the state — of regional and local government.
Historical geography	The long-term historical and prognostic trends in regional development of the European countries.	The recent achievements in historical and prognostic studies.
Prognostic geography		The methodology of alternative scenarios applied in prognostic studies.

disciplinary inspiration for Polish geography (Kostrowicki, 1983). It is not yet clear to what extent this new source will transform conventional approaches to physical geography, but it is clear that it provides the best vehicle for stimulating a more human orientation in physical geography.

A second new source of cross-disciplinary inspiration comes from the recent achievements in the social sciences and humanities, especially sociology, social ecology, psychology and cultural anthropology. This is having an important influence on the methodological reconstruction of social geography and on the creation of a new field, the geography of culture.

New trends in the economic and managerial sciences are also influencing Polish geography. Particularly important is the absorption of the theoretical ideas of J. Kornai (1980) into the reconstruction of economic geography. Similarly, a new shape for Polish political geography should reflect developments in politology and administrative science (Kuklinski, 1983b).

Lastly, there is the inspiration for geography of developments in history and futurology. Both historical and prognostic approaches have well-established traditions in Polish geography but recent developments in the field of long-term (past and future) analysis have yet to be absorbed. There is currently a crisis in futurology, a consequence of its unpleasant defeat in the 1970s. But that was not fatal, and futurology and long-term prognosis, with all their weaknesses, will remain a necessary avant garde with policy and planning. As such, both historical and prognostic geography are promising fields for expansion of research activity.

As with the list of new questions, this list of the sources of new cross-disciplinary inspiration could be expanded by other geographers. In all cases, they would be identifying elements of the second scientific revolution in Polish geography, which is currently creating new interrelationships with other disciplines.

Polarized Development in the 1980s

Changes in Polish geography are not occurring with the same speed and efficiency in the various parts of the discipline. Once again, polarised development is taking place, this time with the launch of paradigm III. The latter is still in a very fluid state, however, for paradigm II has not yet died and is fighting to retain its place for the mainstream position in Polish geography. The trends involved in this fight are outlined in Table 7.2, which can be compared with Table 7.1.

Both Table 7.1 and Table 7.2 are over-simplifications, however. I am not arguing that the traditional growth poles of Polish geography (geomorphology, geography of agriculture, settlement geography) will not flourish, for example. What I am presenting in Table 7.2 is a hypothesis about the new intellectual growth poles, whose precise shape is yet to be determined.

A New Look at the History of Geography of Polonia Restituta

The materials and arguments presented here and elsewhere (Kuklinski, 1983b and 1983d) lead to the following tentative conclusions about Polish geography in the period 1918-2000. Six periods can be identified:

I   1918-34: the construction of paradigm I[3]
II 1934-49: the domination of paradigm I
III 1949-60: the first scientific revolution — the annihilation of paradigm I and the construction of paradigm II
IV 1960-76: the domination of paradigm II
V   1976-85: the second scientific revolution — the accelerated decline of paradigm II and the construction of paradigm III
VI 1985-2000: the domination of paradigm III.

Polish Geography; Polish Society; History of Poland

No academic discipline occupies a social, economic, political and cultural vacuum. This is a universal rule with special importance and validity for Poland; it is for this reason that the development of Polish geography has been analysed here in the context of endogenous and exogenous influences. The paradigms and revolutions that I have identified are, as I have stressed, personal judgements; for others, they may form hypotheses to be tested against the literary record.

To appreciate the contextual approach developed here, readers outside Poland may need additional knowledge of the country's society, economy and history (see Szczepanski, 1970; Gieysztor, Kieniewicz, Rostworowski, Tazbir and Wereszycki, 1979; Bromke, 1982). With this, they will be better able to evaluate the discussion. In addition, it is hoped that the chapter will be seen not only as a contribution to the history of Polish geography but also as a catalyst for the formulation of questions to be addressed to the historical experience in other countries.

Notes

1. Among the contents of that issue are papers by: K. Dziewonski ('Geography of settlement and population in Poland 1945-1982'); J. Kostrowicki ('Polish agricultural geography: mainsprings of development'); and A. Kotarba, S. Kusarski and L. Starkel ('Mechanisms in the development of Polish geomorphology').

2. An analogy is provided by Galbraith (1955, pp. 44 and 48) in his references to 'the issue of automacity as against the guided economy' and 'the rejection of the notion of automatic good performance by the economy'.

3. This is paradigm I of Polonia Restituta – Poland Reborn; there is a long history of pre-1918 Polish geography not referred to here.

References

Works with a Polish text are indicated by an asterisk.

Bromke, A. (1982) *The Revival of Political Idealism in Poland*, Canadian Slavonic Papers, Canadian Association of Slavists.

*Cackowski, Z. (1972) 'Problemy teorii odkrycia naukowegoo' (Problems of the Theory of Scientific Discovery), *Studia Filozoficzne* (Philosophical Studies), 7-8, 127-39.

Chojnicki, Z. (1983) 'Methodological Problems in Regional Science' in A. Kuklinski and J.G. Labooy (eds.), *Regional Policy – The Crossroads of Theory and Practice,* UNRISD-Mouton Regional Planning Series, vol. 11, Mouton, Amsterdam, 25-45.

Dziewonski, K. (1975) 'Research for Physical Planning in Poland 1944-1974', *Geographia Polonica*, 32, 5-23.

*Dziewonski, K. (1983) 'Geografia osadnictwa i ludnosci w Polsce 1945-1982' (Geography of Settlement and Population in Poland 1945-1982. Mechanisms of Development), *Polish Geographical Review, 55*, 3.

——— (1984) 'National Settlement Systems in Comparative Studies: 1976-1984', *Geographia Polonica, 50.*

Galbraith, J.K. (1955) *Economics and the Art of Controversy*, Rutgers University Press, New Brunswick, NJ.

Gieysztor, A., Kieniewicz, S., Rostworowski, E., Tazbir, J. and Wereszycki, H. (1979) *History of Poland*, Polish Scientific Publishers, Warsaw.

Gilewska, S., Klimkowa, M. and Starkel, L. (1982) 'The 1:500 000 Geomorphological Map of Poland', *Geographia Polonica, 48,* 7-25.

Institute of Geography, Polish Academy of Sciences (1959) *Problems of Applied Geography*, Proceedings of the Anglo-Polish Seminar,

Nieborow, September 15-18, Geographical Studies, 25.

Jankowski, W. (1982) 'Polish Experiences in Land Use Mapping', *Geographia Polonica, 48,* 59-71.

Jedraszko, A. (1983) 'Physical Planning in Poland: The Changing Fate of the Doctrine' in A. Kuklinski and J.G. Lambooy (eds.), *Regional Policy – The Crossroads of Theory and Practice,* 435-57.

Johnston, R.J. (1979) *Geography and Geographers. Anglo-American Human Geography since 1945,* Edward Arnold, London.

Komorowski, S.M. (1982) *Regional Studies: The Geneva Programme,* UNCRD Working Paper No. 82-1, United Nations Centre for Regional Development, Nagoya.

Kornai, J. (1980) *The Economics of Shortage, cols. A and B,* North-Holland Publishing Company, Amsterdam.

*Kostrowicki, A.S. (1983) 'Dialog geografii z ekologia' (The Dialogue of Geography and Ecology), *Polish Geographical Review, 55,* 3.

*Kostrowicki, J. (1981) 'XXIV Miedzynarodowy Kongres Geograficzny, jego problematyka i wyniki a pozycja geografii polskiej' (XXIV International Geographical Congress – Problems and Results versus the Position of Polish Geography), *Polish Geographical Review, 53,* 3, 447-73.

*—— (1983) 'Polska geografia rolnictwa. Mechanizmy rozwoju' (Polish Agricultural Geography. Mainsprings of Development), *Polish Geographical Review, 53,* 3.

—— (1984) 'Types of Agriculture in Europe. A Preliminary Outline', *Geographie Polonica, 50.*

*Kotarba, A., Kozarski, S. and Starkel, L. (1983) 'Mechanizmy rozwoju polskiej geomorfologii' (Mechanisms in the Development of Polish Geomorphology), *Polish Geographical Review, 55,* 3.

Kuhn, T.S. (1962) *The Structure of Scientific Revolutions. International Encyclopedia of Unified Science,* The University of Chicago Press, Chicago.

Kuklinski, A. (1966) 'Research Activity of the Committee for Space Economy and Regional Planning' in J.C. Fisher (ed.) *City and Regional Planning in Poland,* Cornell University Press, Ithaca, New York, 389-405.

—— (ed.) (1972) *Growth Poles and Growth Centres in Regional Planning,* UNRISD-Mouton Regional Planning Series, vol. 5, Mouton, Paris-The Hague.

—— (1982) *Space Economy and Regional Studies in Poland,* United Nations Centre for Regional Development, Nagoya.

.*—— (ed.) (1983a) 'Diagnoza stany gospodarki przestrzennej Polski.

Wstepne wyniki badan' (Diagnosis of the State of Space Economy in Poland. Preliminary Results), *Bulletin of the Committee for Space Economy and Regional Planning*, 123, Polish Academy of Sciences, Warsaw.

* —— (1983b) 'Mechanizmy rozwoju geografii polskiej w latach 1945-1982. Artykul dyskusyjny' (Polish Geography-Mechanisms of Development in the Years 1945-1982), *Polish Geographical Review, 55,* 3.

—— (1983c) 'Polish Space in the Eighties', *GeoJournal,* 7, 1, 80-3.

—— (1983d) 'Studies in the History of Polish Geography: Observations and Reflections', *GeoJournal.*

—— (ed.) (1984) 'Regional Studies in Poland – Experiences and Prospects', Studia Regionalia (Polish Academy of Sciences, Warsaw).

Leszczycki, S. (1972) 'Geography in Poland: Main Trends and Features', *Geographia Polonica, 22.*

Mikulinski, S.P. and Jaroszewski, M.G. (1969) *Naukowiedienije. Problemy i issledowanija. Naucznoje tworczestwo,* Izd. Nauka, Moscow.

Oeconomica Polona, Journal of the Economic Committee of the Polish Academy of Sciences and of the Polish Economic Society (10 volumes published in 1973-83).

Polish Academy of Sciences (1982) *Problemy polskiej przestrzeni* (Problems of Polish Space), Bulletin of the Committee for Space Economy and Regional Planning, 118, Warsaw.

Szczepanski, J. (1970) *Polish Society*, Random House, New York.

Wolfe, M. (1981) *Elusive Development,* UN Research Institute for Social Development and UN Economic Commission for Latin America, Statistical Publishing House, Budapest.

Ziman, J. (1968) *Public Knowledge,* Cambridge University Press, Cambridge.

8 THE GERMAN-SPEAKING COUNTRIES

Elisabeth Lichtenberger*

The development of geography in the German-speaking countries since 1945 is outlined here in the context of political systems, institutional structures and the organisation of research. (For other reviews, see Troll, 1949; van Valkenburg, 1951; Hajdu, 1968; James, 1972; Manshard, 1976; Hellen, 1978.) It is well established that the nature of scientific research is conditioned by political systems and by investment in human and physical capital. Thus, in the present context, the two Germanies plus Austria and Switzerland provide a 'field laboratory' for investigating the impact of those political systems (Lichtenberger, 1979), particularly the institutional growth that has taken place since 1945 (Lichtenberger, 1981). Within this framework, the research process can be analysed according to both the basic ideologies and philosophies (Lichtenberger, 1978) and the empirical subject-matter and achievements (Blotevogel and Heineberg, 1976-81).

The Impact of Political Systems

The FRG and the GDR

The practice of geography differs between these two countries as a consequence of their respective allocation to the American and Russian political hemispheres. As a matter of course, geographers in the FRG rely mainly on British and American academic communication and information systems, whereas those in the GDR are linked to Russian sources.

The institutional structures vary also. In the FRG (and also in Austria and Switzerland) research takes place within a liberal, *laissez-faire* context, influenced only indirectly through the national research foundations. In the GDR, on the other hand, research is centrally directed and depends directly on the resolutions of the *Sozialistische Einheitspartei Deutschlands* (SED) Party Conventions. Here research

*This abridged and rewritten version of the original text has been prepared by R.J. Johnston. Because of the pressing deadline it was impossible to include the author's comments and revisions.

is conducted not in the universities – which in western countries unite research and teaching – but in the Academy of Sciences; some university geographers have links with the Academy, but it is a rather exclusive institution containing a circle of scholars with security of tenure. This closed structure has provided limits to the expansion of geographic research (in breadth and volume) whereas in the FRG the growth of universities has fuelled a diversification of research activity, along the American and British lines. In the 1960s and 1970s the foundation of new universities provided employment opportunities for geographers, and the total teaching staff is now five times its number in the 1950s (see Table 8.1). For their counterparts in the GDR, however, some institutes were abolished during the third university reform of 1967-9. Reflecting this difference, the number of journals published in the FRG has multiplied (see Harris, 1980) but there are only a few in the GDR, each with a specific task and supervised by the Geographical Society (itself linked to the Academy of Sciences since 1969).

Political doctrines, and their representation in the scholarly vernacular, have produced a growing divergence between geographers in the two countries. Even basic models, such as Christaller's central-place theory, have different interpretations: in the FRG it is implicitly a consumer-orientated supply model, whereas in the GDR it has been adapted to the ideology of a production-orientated society. In general, GDR geography is characterised by neo-positivist and neo-Marxist terminology, whereas west of the Iron Curtain there is an ideology of self-identification and behaviourism.

Research orientations differ substantially between the two countries. In the GDR, almost all of the research is conducted within the country, with the Academy of Sciences receiving public commissions and allocating them to the various institutes (which have a total research staff of about 80 geographers). There is no equivalent of these institutes in the FRG, where much research is undertaken abroad; within the country, as well as undertaking basic research, geographers can participate in applied research opportunities provided by private and semi-official institutions, and also by state bodies. And geographers there can get involved in open and candid critical discussions (as in the session at the meeting of geographers chaired by Peter Scholler at Kiel in 1969, when time was given for a discussion of neo-Marxist concepts introduced by students). In the GDR, discussion is somewhat sterilised by the dogma of historical-dialectical materialism.

The GDR

A centralised research structure has been established in the GDR, under USSR influence, in several stages In 1950 H. Sanke (1980), who had been re-trained in Moscow after the Second World War, was appointed professor in Berlin, leading to the replacement of the former discipline of human geography by Marxist-Leninist economic geography. Nevertheless, there was direct communication between geographers in the two Germanies during the 1950s, and they shared in projects on the geomorphological and ecological structures and central-place systems of all Germany. Major changes came after the building of the Berlin Wall in 1961, and the *Geographentag* at Bochum in 1965 was the last attended by geographers from the GDR. Since then, there has been hardly any communication between the two groups of geographers and developments in the GDR have focused on its own Geographical Society (founded by E. Neef in 1953) and its own publication, the *Geographische Berichte*.

The third university reform of 1967-9 produced very serious effects for geography. Three geographical institutes were closed, at Rostock, at Jena and at Leipzig (where Richtofen, Ratzel and Partsch had all worked). Only five geographical institutes remained, at the Humboldt University of Berlin, Greifswald, Halle, the Teachers' Training College at Potsdam (where geography is taught with physical education) and at Dresden where geography is incorporated with geodesy. An Institute of Geography and Geoecology within the Academy of Sciences was established by E. Lehmann, however, in the premises of the former Deutsches Institut für Landerkunde.

According to the USSR model, geographical research was reorganised in a series of public commissions operating within a central research plan (Haase, 1977; Lüdemann, 1980; Zimm, 1982). The contracting partners are the State Planning Boards, the Bezirke (district) and East Berlin planning authorities, and the Ministry of Economy and Technology. The main tasks are to participate in: (1) the formation of an effective territorial structure of the reproduction process of society; and (2) the unkeep and improvement of natural resources for both human life and production.

These two main tasks focus research in two areas. The first is *ecological landscape research* (Neef, 1967), which involves geo-ecological classification and the economic evaluation of natural resources. The detailed geological/pedological mapping and establishment of hydrological balances forms the basis for the taxation of state farms. *Territorial research* (Grimm, 1979; Mohs, 1980; Krönert, 1981), the second focus, involves economic geography as an integral part of the interdisciplinary research that provides basic data and prognostication instruments for the planning of social organisation (Schmidt-Renner, 1970; Schmidt-Renner *et al.*, 1973). Methods of regional science are used in this work, and Haggett's book *Geography – A Modern Synthesis*, translated in the USSR, is a basic source. Geographers deal in particular with the macro-structure of the settlement system, with urbanisation and the problems of metropolitan areas and conurbations, and with the provision of infrastructure and communication networks.

Geographers are involved in other tasks, as in the preparation of scenarios for multiple land-use schemes. These scenarios for future use and the simulation of development trends seek to optimise land-use distributions within given social constraints, such as economy of expenditure, social effectiveness, protection of natural surroundings and the preservation and regeneration of natural resources. Methodologically, stress is on the invariant structural elements, with little attention to the dynamics of geosystems.

The main contacts for GDR geographers are with their Russian counterparts. There are intensive interchanges at congresses and symposia, and Soviet authors publish frequently in GDR journals; GDR geographers have participated in the analysis of Russian remote-sensing data. Many research findings are kept secret, however, and this prevents public discussion. Research abroad is very difficult.

Switzerland and Austria

These two small countries have much in common with regard to the practice of geography (Lichtenberger, 1975; Geography in Switzerland, 1980). Each has a small number of institutes and staff, necessarily leading to the monopolisation of certain fields by various experts (e.g. E. Winkler and H. Carol on theory and H. Bösch on economic geography, in Switzerland; E. Arnberger on cartography, J. Fink in quaternary research and H. Kinzl on demography, in Austria). Both, too, have experienced an emigration of scholars to the FRG.

Research is focused on the home countries, except for that associated with the impressive amount of development aid provided by Switzerland, and there is substantial participation in applied research. The latter has promoted problem-orientated, co-operative work (notably on regional policy) at the expense of historical explanation and analytical thinking; work in the traditional geographical subdisciplines has been replaced by a concentration on interdisciplinary problem areas. This is reflected in the strength of the Vienna school of urban geography (active since Hugo Hassinger's time), and its fundamental research on central places has become an integral element in the regional planning process (Hassinger, 1938; Bobek and Fesl, 1978, 1983). The relationship with regional planning has provided an opportunity for geography graduates to enter the higher strata of the civil service.

Much research in these two Alpine states not surprisingly focuses on various aspects of the mountain environment. Glaciology has attracted geographical interest (in some cases in co-operation with the Alpine Clubs), for example, and in Switzerland half of all the climatologists are geographers by origin. Farming and tourism in these areas, along with issues relating to population, settlement and economy, have been research themes of long standing (Kinzl, 1940; Fliri, 1975; Messerli *et al.*, 1978; Keller, 1979).

The FRG: The Consequences of Institutional Growth

Based on policies aimed at equal educational chances for all population groups, an educational revolution took place in the FRG in the 1960s and 1970s, in which geography participated. New universities were founded on the principle of one institution per million inhabitants in a nucleated region; the number of geography institutes doubled and the staff quintupled (Table 8.1). The great increase of students came in the 1970s, however, when the expanding supply of teachers had to be halted because of the economic recession; indeed, the number of staff has been reduced at some institutes. The gap between the supply of and demand for courses widened (Table 8.2), and teaching routines had to change – freedom of choice of courses was restricted for students, and teaching methods for staff had to change (the average university teacher in Germany has a teaching load twice that of his British counterpart, and handles three times the number of students; the time available for research is consequently less). The demand for teachers of geography is declining, so that the nature of course offerings has been engineered, with a greater orientation to 'pure' geographers (those studying for the MA and MSc); nevertheless, there are fears that the numbers of those who are either unemployed or employed in jobs for which they are overqualified will increase.

During the period of growth, university and school geography were brought into close contact. University staff were involved in curriculum development, following the research initiated by Geipel (1978). The interest of commercial publishers in the large market prompted

Table 8.1: The Development of the Institutes of Geography in the FRG, GDR, Austria and Switzerland, 1955/6 and 1979/80

	FRG		GDR		Austria		Switzerland[a]	
	1955	1980	1955	1980	1955	1980	1955	1980
	Number of institutes							
	30	60	8	5(4)	4	6	6	5
	Number of teaching staff							
Full professor	31	202	7	15	5	14	7	7
Associate professor	30	18	13	7	5	3	4	6
Tenured post	–	115	–	–	–	–	–	–
Lecturers	5	22[b]	9	20	10	21	2	26
Assistants	43	300	49	19	6	44	11	51
Others	7	11	9	13	–	2	1	11
Total	116	668[c]	87	74	26	84	25	101

Notes: a. German-speaking area.
         b. Without part-time lecturers.
         c. Without undergraduate assistants.
Source: *Geographisches Taschenbuch* (1957/8, 1980/1) *Orbis Geographicus* (1980/4).

Table 8.2: The Number of Students of Geography, and the Number of 'Pure' MA Students, 1978/9 to 1981/2

	Total	MA
1978/9	25,800	3,000
1979/80	25,800	3,200
1980/1	29,500	4,700
1981/2	34,000	5,800

Source: A. Mahr, *Geographische Rundschau*, 35,5, 1983.

the production of geography textbooks: these reached high standards, in part because of competition generated by the federal structure of the education system.

For research publications, there has been no increase in the number of national journals, although the traditional *Geographische Zeitschrift* was revived and set alongside *Erdkunde*, founded by Carl Troll. Institutional particularism has flourished, however, and at present there are six traditional journals published by geographical societies plus 55 series issued by university institutes. (Harris, 1980, lists only 36.) Unfortunately, therefore, no central information and discussion fora have emerged. Bibliographical filters are also lacking, apart from the *Geolit* series published by Westermann and concerned only with books. And there is little inclination among German geographers to popularise their work, with no comparable journal to the *Geographical Magazine* published in Britain.

There are two nationally important organisational structures for contemporary geography in the FRG. The *Dachverband* (Federation) of geographers combines the associations of geographers in universities and schools, plus the 'professional' geographers; the *Deutscher Geographentag* is a well-organized biennial meeting at which current research is discussed. For the prosecution of research, there is much reliance on support from the *Deutsche Forschungsgemeinschaft* (DFG; the German Research Foundation). Specific boards within this allocate finance to projects, on the basis of internal evaluations by the disciplines. The export-orientated economic strategy of the FRG is reflected in the willingness of opinion-leaders within the DFG to support research abroad on a large scale.

The Succession of Ideologies

Classical German geography is often labelled 'landscape geography', a term which does an injustice to the pluralistic structure and which applies only to the subset of 'core' subjects — historical settlement and cultural geography, and genetic geomorphology (Lichtenberger, 1978). Both of these cores are based on 'ideal types' for their focus of attention, and are organised around theories and hypotheses centred on the historical succession of generations of forms. They lack neighbouring systematic disciplines, unlike the 'frame' subjects — such as climatological geography, hydrogeography, and population and social geograpy.

Until the end of the 1950s, the normal career pattern for geographers involved a transition from an early specialisation in geomorphology to a focus on settlement geography. The result was a transfer of modes of scientific reasoning and research methodology from geomorphology to historical landscape analysis; genetic regularities were sought, and the hermeneutic approaches were unknown, as illustrated by work on settlement patterns, field systems and agricultural geography. Methodologically, it led to high standards of field techniques, a predominance of micro-scale studies with the diligent compilation of original data in the field, excellent standards in the analysis and inter-pretation of aerial photographs, and high quality work in cartography. (Cartographic documentation at the micro-scale remains a strong point of German geographical research.)

This historical-morphological approach to the study of landscape changed to a functionalist approach after the Second World War, with the introduction of (qualitative) exogenous explanatory variables to the models, such as climatic change and social behaviour. Thus the formulation of theories at the macro-level was combined with ex-tremely detailed micro-scale analyses with landscape observations of particular indicators providing the means of studying various pheno-mena. During the 1960s the tie with the landscape was broken, with growth of the study of geomorphological processes (morphodynamics) on the one hand and the behaviour of social groups on the other.

By the end of the 1960s a pluralistic structure of the discipline had emerged, even in the 'core' (Lichtenberger, 1978). This was being con-fronted, especially in the 'frame', with the analytical developments in Anglo-Saxon geography. The infiltration of these was slow, however, for a variety of reasons. Firstly, the importance of overseas research, and the mistrust of official statistics which it engendered, provided a considerable barrier to methodological changes. Secondly, the editing of journals and the production of textbooks is controlled by the estab-lished generation of professors. Finally, the inertia of research methods developed in climatic geomorphology and historical settlement geo-graphy proved difficult to overthrow.

Almost simultaneously with the development of the new analytical scientific approaches came the demands, largely from students, at the 1969 *Deutscher Geographentag* for a Marxist approach to geography. The concepts and the socio-political context of these demands differ substantially from those of Anglo-American 'radical geography', how-ever, for ideas developed within critical theory have entered all levels of socio-political argument in the German-speaking countries. The goal has been to develop an 'engaged geography' with a pluralistic ideolog-ical outlook, involved in decision-orientated approaches. Only human geography has been substantially influenced, with physical geographers finding the discussion of paradigms irrelevant.

Research Trends in the FRG, as Examplified in the Deutsche Geo¬
graphentage

The biennial meetings of German geographers, where current research
findings are presented, provide a series of examples of the changing
foci of work. Each meeting has certain unique features, reflecting the
interests of the hosts (e.g. 1955, Hamburg, overseas research; 1971,
Erlangen, oriental studies; 1975, Innsbruck, high mountain environments). But there are common features, such as strict programme
planning and the selection of papers to be read, and published in the
excellently edited series of Proceedings which provides a basic docu-
ment on changing German geography.

Despite the institutional growth, the number of papers presented did
not exceed 40 until 1977, except for 52 in 1969 and 56 in 1975. There
was a complete reorganisation in 1979 (Göttingen), with main sessions,
sub-sessions and study groups for current research problems; 132 papers
were submitted.

A classification of the papers presented (Table 8.3) indicates the
relative importance of both human geography and regional geography
overseas. The general trends over time are shown in Figure 8.1 (these
have been 'smoothed' to remove the influence of local interests). Thus
physical geography is currently experiencing a resurgence of interest,
following a decline in the 1960s (Ahnert, 1978), whereas regional geography has been losing ground to systematic studies in recent years.

The main features of the various *Geographentage* are:

(1) The first meetings, in 1948 and 1951, were dominated by the
     'papers' of certain research traditions, such as Büdel on clima-
     togenetic geomorphology, Troll on vegetation geography, Christ-
     aller on central places, Bobek on social geography and Flohn on
     synoptic climatology. Other flourishing areas were quaternary
     research (Stäblein, 1980), historical-cultural landscapes (Jäger,
     1972; Uhlig and Lienau, eds., 1974), and agricultural geography.
(2) In the 1960s work on historical-cultural landscapes declined but
     the Munich school of social geography, and its offshoot studying
     educational structures, flourished. Urban geography climaxed at
     the 1967 Bad Godesberg meeting (Schöller, 1973). There were
     several gaps, however, including population and industrial geo-
     graphy, the geography of tourism and pedology.
(3) The growth of analytical work, introduced by Bartels (1970) at
     the Kiel meeting in 1969, was finalised by the constitution in
     1974 of a small study group in quantitative and theoretical geo-

Table 8.3: Topical Structure (in percentages) of the Papers Published in the Proceedings of the 'Deutschen Geographentage', 1948-83

	Systematic geography	Regional geography FRG	Regional geography Europe	Regional geography overseas	Total
Theories, methods	1.8	—	—	—	1.8
Physical geography	9.9	5.3	3.4	11.1	29.6
Human geography	27.4	12.1	9.7	19.4	68.5
Total	39.6	17.4	13.1	30.5	100.0

n= 922[a]

Note: a. Without papers on didactics.
Source: Proceedings of the 'Deutschen Geographentage', 1948-83.

Figure 8.1: (a) The Trends in Physical and Human Geography, 1948-83.

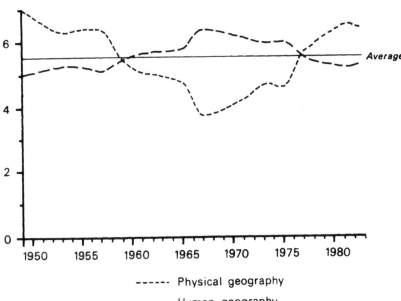

------ Physical geography
— — — Human geography

Figure 8.1: (b) Regional Geography, Theory and Methodology, 1948-83

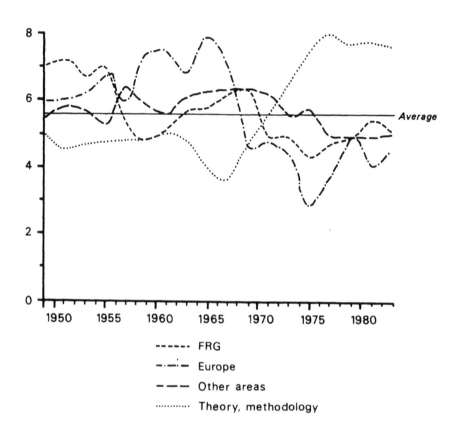

graphy (Giese, 1981).

(4)   During the 1970s new research fields were introduced as results of external contacts, including landscape ecology and medical geography (1973), high mountain environments, anthropogenic morphodynamics and the world's carrying capacity related to population growth (1975; see Penck, 1925 and Rathjens, 1979). Developments in physical geography have been largely concerned with resource management (Hard, 1976).

(5)   The provision by the established generation of scholars of subdisciplinary surveys — for example, for climatic geomorphology (Hagedorn and Thomas, 1980), rural settlement studies (Nitz, 1974), urban geography (Lichtenberger, 1983) and industrial geography (Hottes and Hamilton, 1980) — ended with the Göttingen meeting in 1979.

Developments in the various subdisciplines have been undertaken with different degrees of contact with neighbouring disciplines – as already noted with regard to the 'core' and 'frame' components of geography. Physical geographers had a much longer tradition of such contact, for example, and so were relatively untouched by the analytical developments derived from Anglo-Saxon geography. (Physical geography is best represented in the various academies, as follows: earth science, Mainz; geomorphology, Göttingen and Munich; climatic research, Düsseldorf; glaciology, Munich; alpine geography, Bern; regional research, Vienna.)[1] Human geographers broke their ties with history, however, and formed new relationships with the social sciences, adopting their theories, problems and research strategies; nevertheless, research into historical-cultural landscapes has increased since 1979.

A study of the publications of a selection of established physical geographers aged 40-50 documents the intra- and interdisciplinary links that have developed (Figure 8.2). Nearly one-fifth of their publications were in human geography, indicating that it is still common to shift from physical to human interests. Interdisciplinarity was highest in pedology, as was the flow of material to English-language journals. Geomorphology remained the 'core' subdiscipline, and its most important focus was climato-genetical studies (Bremer, 1983); close links with Quaternary research (tracing back to the work of A. Penck) have been maintained, with work on Arctic (Stäblein, 1983), arid (Central Asian: Hövermann and Wang Ying, 1982; North African: Mensching, 1974, 1982), and humid tropical environments. Periglacial research has occupied a relatively independent position, and great progress has been made in Karst studies (Sweeting and Pfeffer, eds., 1976). Research achievements include a standardised scheme for geomorphological mapping at the 1:25,000 and 1:100,000 scales aided by a high-priority DFG-award (Barsch and Liedtke, 1980). Analytical geomorphology has emerged only recently, as in the study of soil erosion processes (Richter and Negendank, 1977; Leser, 1980).

In the 'frame' areas, developments are very dependent on the links with neighbouring disciplines. There are, for example, strong geo-branches of botany and pedology. The small group of workers in vegetation geography, led by Troll, has participated in research on quaternary problems, cultural history (e.g. the formation of steppes and moorlands), the savanna problem, deforestation in the Mediterranean and current desertification (Mensching and Ibrahim, 1977). Such geography had a less firm base, having been identified largely as an auxiliary to quaternary research (Semmel, 1977); climatology, through the

Figure 8.2: Interdisciplinary Nexus of Subdisciplines in Physical Geography and Publications in Foreign Languages

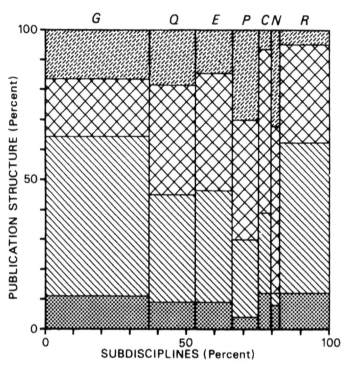

G   Geomorphology
Q   Quaternary research
E   Ecology
P   Pedology
C   Climatology & Hydrology
N   Neighbouring disciplines
R   Regional Geography, etc.

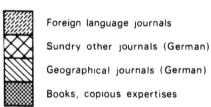

Foreign language journals

Sundry other journals (German)

Geographical journals (German)

Books, copious expertises

Source: List of publications of Noted German physical geographers in the 40 to 50 age group (n = 22).

influence of the meteorologist Flohn, has created several subdisciplines such as agricultural, urban, terrain, tropical and applied climatology (Eriksen, 1975); and hydrological geography developed its reputation through interdisciplinary participation during the International Hydrological Decade (Keller, 1961, 1974; Zötl, 1974). Two new areas of work, in which neighbouring disciplines have failed to develop geo-branches, are zoogeography (Müller, 1974) and medical geography (Jusatz, ed., 1967-80).

In human geography, changes were related not only to links with other disciplines but also to changes in society as a whole. Research into historical-cultural landscapes, developed in the period 1930-60, was a product of the German 'educated middle-class' interests; its methods included the study of documents, archaeological excavations, soil analyses, mapping of plant communities and the establishment of chronologies (Jäger, 1981). Social geography, focused on the Munich school, developed late in the 1960s, influenced by American and British sociology; its leaders were Bobek in Vienna and Hartke in Munich (Thomale, 1972). The Munich school focused on micro-behavioural studies (Ruppert, 1983), whereas Bobek (1962) and his co-workers (e.g. Müller, 1983) concentrated on macro-issues, such as stages of economic development and the theory of rent capitalism. Their work, which has no counterpart in the Anglo-American tradition, has contributed much to the study of cities in other cultures (e.g. Dettmann, 1970; Hofmeister, 1971; Manshard, 1977; Borsdorf, 1978; Bähr and Mertins, 1981; Borsdorf and Wilhelmy, 1984).

Central-place theory occupied a major position after Christaller's exposition at the 1951 *Geographentag*. It was adapted to industrial environments, intra-city systems, to incorporate stochastic elements, and to potential uses in planning. More recently, research interests have diverged to include topics such as the office agglomeration and Third World periodic markets. Two centres of urban research have emerged, one, led by Schöller, at Münster/Bochum, and the other at Vienna. The work clearly illustrates the dichotomy of research at home and abroad, the complete abandonment of the landscape concept, and the widening spectrum of methods.

Among the other components of human geography, social geography had already forged strong links with neighbouring disciplines, and these were furthered during the 1970s. Their tradition of empirical micro-scale research gained geographers a place in interdisciplinary teams studying such topics as mobility, segregation and guestworkers (Bartels, 1968; Hermanns, Lienau and Weber, 1979). Within economic geography, the links were initially much weaker. Recently, concepts derived from political economy have been incorporated within new geography texts (Schätzl, 1981), and industrial geographers have shifted their focus from a physiognomical-morphological perspective to the use of behaviourist concepts. Perception research has been developed by

the Munich school also (Geipel and Pohl, 1983), though links with both
social psychology and management science are as yet weak.

Human geographers have brought two sets of technical skills to
these developing links. The first are concerned with the use of aerial
photographs and remotely-sensed data in cartography (Schneider, 1974;
Gierloff-Emden, 1977). The second are the body of mathematical and
statistical methods (Bahrenberg and Streit, eds., 1981). As yet, there
has been an emphasis with these methods on logical rigour rather than
a practical usage, and the framework of mathematical model-building
has been to some extent 'art pour l'art' (Bahrenberg, Fischer and
Nijkamp, eds., 1983).

Research at Home and Abroad

Foreign-based research has claimed a great deal of German geographers'
attention, because of its position in the discipline's career norms and
the support provided by funding bodies. A much greater proportion
of German (80 per cent) than American (c. 10 per cent) geographers do
field-work abroad (Mikesell, ed., 1973). Data on the projects abroad
sponsored by the DFG clearly indicate the research politics that have
promoted a 'luxurious' export of scientific expertise. (Both the Volks-
wagen and Thyssen Research Foundations follow similar strategies to
the DFG's.) During the period 1976-81, the DFG sponsored as many
projects in the Near East as in the FRG (13.6 per cent each), and there
were as many in the Americas as in the whole of Western Europe
(Table 8.4).

Three different institutional types of foreign-area research can be
identified. *Individual research* predominates, and almost every large
geographical institute has its own expert on Africa and Latin America.
There are some *durable regional schools*, however, such as that led by
E. Wirth (Erlangen), focusing on the Orient, and that led by G. Sandner
(Hamburg), which focuses on Central America. Finally, there are the
DFG's centres, with chairs in certain institutions earmarked for specialists on specific areas. There are few institutes for foreign area studies
(South-eastern Europe, Munich; North America, Berlin; Africa, Ham-
burg; South-eastern Asia, Heidelberg; Eastern Asia, Berlin) but in
general geography is not integrated in their work.

Within Europe, there is relatively little contact between German and
French geography, and each has its own sphere of influence; the French
dominate in South-western Europe, Spain and Italy, whereas the

Table 8.4: Individual Projects Sponsored by the German Research Foundation

Research fields	1961-1975[a] Number of projects	%	1976-1981[b] Number of projects	%
1. *FRG, GDR*	237	16.5	46	13.6
2. *Europe*				
Western Europe	318	22.1	39	11.6
Eastern Europe	36	2.5	8	2.4
Soviet Union	15	1.0	–	–
Total	606	42.1	93	27.6
3. *Overseas*				
Islamic Orient	132	9.2	46	13.6
African south of the Sahara	140	9.8	31	9.2
Indian subcontinent	64	4.5	8	2.4
South-east Asia	48	3.3	22	6.5
East Asia	36	2.5	13	3.9
Australia	28	2.0	7	2.1
Latin America	138	9.6	40	11.8
North America	90	6.3	35	10.4
Total	676	47.2	202	59.9
4. *Systematic geography*	73	5.1	28	8.3
Others	80	5.6	14	4.2
Total number of applications for DFG grants	1.435	100.0	337	100.0

Sources: a) Far-Hollender, U. and Ehlers, E. (1977): 'Geographische Forschungen in der BRD und die Deutsche Forschungsgemeinschaft 1960-1975' in E. Ehlers and E. Meynen (eds.) *Geographisches Taschenbuch 1977/8*, Steiner, Wiesbaden, 241-53.
b) *Jahresberichte der Deutschen Forschungsgemeinschaft — Programme und Projekte*, Bonn-Bad Godesberg, 1976-81.

Germans have a long tradition of research in Eastern and South-eastern Europe and the Near East (Karger and Sperling, 1980; Stadelbauer, 1983), and better contacts with North-western Europe. In recent decades micro-scale research in Eastern Europe has been almost impossible, except in Yugoslavia, but sources within the FRG have provided the needed materials for historical studies (Jäger, 1982). Both the Munich (Ruppert, ed., 1981) and the Vienna schools (Breu, 1970; Leitner, 1982; Lichtenberger, forthcoming) have maintained research links with Yugoslav geographers, and Fink for a long time co-ordinated joint pedological and quaternary research. Further south, Greece has been a domain for German geographers since the nineteenth century, and there have been recent projects relating to guestworkers and remigration.

In the Near East, German research occupies top position with regard to volume, scope and quality of publications (Wirth, 1983). This is certainly the case with historico-geographical studies of settlements (Wissmann, 1968; Huttoroth and Abdulfattah, 1977) and also with the study of the characteristic features of cities there, such as their bazaars (Wirth, 1974-5). Much progress has been achieved in the study of agricultural systems, including changes in the nomadic economy, success and/or failure of agrarian reforms, and problems of irrigation (Jentsch, 1973; Schweizer, 1974; Kopp, 1981; Schulz, 1981; Ehlers, 1982).

As well as detailed research reports, many regional handbooks have been produced (on Turkey, Iran, Syria, Egypt, Afghanistan and Oman, for example) which have no counterparts in other languages. There are also many detailed monographs, notably on the towns of the area. Among the major achievements is the *Tübingen Atlas of the Near East* (DFG, ed., 1978), an interdisciplinary product, comprising 375 maps with accompanying text, in which geographers played a decisive role.

Only a few German geographers have worked in South, South-east and East Asia, but almost all have been outstanding scholars who have greatly influenced the development of the sciences there. They include Lautensach (1945) who worked in Korea, Kolb (1963) on China, Uhlig (ed.) (1983) on South-east Asia, Schöller (1980) on Japan, and also Schwind (1967, 1981) who has been awarded the highest Japanese decoration for scientific achievement for his regional geography of Japan. In Africa, too, there has been relatively little work, with a clear distance-decay effect in the intensity of research. North Africa has always received considerable attention. South of the Sahara, the former German colonies have attracted most work, as illustrated by Manshard's (1974) text. There are German geographers working in most countries, but no distinctive schools have developed. A major project has been The DFG *Atlas of African Studies* in four parts (Northern, Western, Eastern and Southern); each is to comprise 16 thematic maps with accompanying text and key in three languages (Freitag *et al.*, forthcoming).

Research in Latin America has several features in common with work in the Near East. There have been major achievements, for example, in the fields of historical geography – the early state in Mexico as revealed by aerial photographic and archaeological evidence (Tichy, 1979), and the geography of the hacienda (Nickel, 1978), agricultural colonisation, agrarian reform and urban studies. There are also close ties between ecological research in high mountain areas (initiated by Carl Troll) and both quaternary and resources-orientated research. A major co-operative project 'to study the history of man from the beginnings to the present in a defined area in its interrelation with the environment' (Lauer, ed., 1976, 1979) was launched in the Puebla-Tlaxaca area of Mexico by Termer and Pfeifer in 1962; to date, it has involved more than 100 representatives of 14 separate disciplines.

Research in Latin America, unlike that in the Near East, has also focused on applied topics, as in the work of Sandner (1982) in Central America and Kohlhepp (1981) in Brazil. But the value of all foreign-area research can be assessed on a variety of criteria, which include: the transfer of information from the periphery to the core and its processing in German texts; a broadening of scholars' mental horizons; and practical co-operation. The first of these has experienced a major boom since the 1960s, with an increasing focus in regional texts on development issues (Bronger, 1977; Krebs, 1982; Krüger, 1982). The theoretical concepts developed in other disciplines have been tested in a variety of regional contexts.

Alongside this foreign-area research, work within Germany has taken second place, and only the Munich school of social geography – which defines its task as 'a service for society by means of work for practical purposes', placing 'economic geography as a basis for planning' – specialises on work within the FRG. Nevertheless, there have been many major achievements, including the maps and text of the *Atlas of the German Agricultural Landscape* (Otremba, ed., 1962 ff). The most important project of the 1950s and 1960s – organised by the *Zentralauschuss für deutsche Landeskunde* (the Central Board of Regional Research) – concerned the central-place system. Its complete stock-taking (Meynen, ed., 1982) provided the basis for local regional and national planning, the first time that geographical concepts had been so important in planning and administration at all levels.

From this base, research into applied geography has developed. Meynen's *Institut für Landeskunde und Raumforschung* at Bad Godesberg was redesignated after his retirement as the *Bundesforschungsanstalt für Landeskunde und Raumordnung*, and its journals were retitled *Informationen zur Raumentwicklung* and *Forschungen zur Raumentwicklung*. Earlier, in 1950 W. Christaller and P. Gauss had founded an association for applied geography (*Deutscher Verband für angewandte Geografie*), which for a long time developed independently of the university institutes. At present applied geography is represented in all of these institutes, and since the 1960s their micro-work has incorporated an awareness of the political aspects (an awareness largely introduced by a generation that had not suffered the shock of the Nazi era).

The various subfields within geography display different balances between research at home and abroad. Both medical and development geography are largely concerned with work overseas, for example, as are climatic geomorphology, vegetation and agricultural geography. Countering this, work in climatological and hydrogeography, glacio-logical and pedological geography, historical geography and the Munich school of social geography is largely concentrated within the FRG — as are the recent imports from the English-speaking world, relating to theoretical and quantitative geography, perception and allocation. Finally, work in urban, population, industrial and tourist geography displays a balance of research at home and abroad.

Summary

This chapter has reviewed geographical work in the German-speaking countries since 1945 in the context of political systems, institutional structures, the succession of ideologies and research realities. The division of Germany in 1945 produced two very separate research strategies, manifested in the ratio of research effort at home to that overseas and in the relative importance of applied work. Nevertheless, in Austria and Switzerland research at home, including applied research, has been more important than it has in the FRG.

Research foci in the FRG have expanded with each new generation of geographers. In the 1950s a few 'popes' founded a small number of research perspectives. The next generation, their erstwhile assistants, introduced new perspectives and expanded overseas work. These were followed by groups who focused on the analytical, quantitative methods, on applied research and development studies; neo-Marxist concepts have infiltrated most areas. Despite all the changes and the growing links with other disciplines (in physical before human geography), the importance given to micro-scale empirical studies and excellence in cartography remains.

Internationally, Germans are no longer simply importers of ideas from the English-speaking world. They are now publishing in the English journals and having their works translated. And there has been no need to adopt the American and British radical geography; the German-speaking countries have their own genuine product in this area, which is potentially exportable.

Appendix

Sources

(1) proceedings of the 'Deutsche Geographentage' (i.e. biennial meetings of German geographers) 1948-1983,

(2) lists of publication of 22 German-speaking professors of physical geography,

(3) reports of the 'Deutsche Forschungsgemeinschaft' (DFG, i.e. German research foundation),

(4) information given by many colleagues in the FRG, for which I am very grateful.

Journals in English Published in the FRG

Geoforum The international multi-disciplinary journal for the rapid publication of research results and critical review articles in the physical, human and regional geosciences. Pergamon Press, Oxford-New York-Paris-Frankfurt. Vol. 1, 1970.

GeoJournal International journal for physical biological and human geosciences and their application in environmental planning and ecology. Akademische Verlagsgesellschaft, Wiesbaden. Vol. 1, 1977.

Catena An interdisciplinary journal of pedology-hydrology-geomorphology. Verlag Lenz, Gliessen. Vol. 1, 1973.

Catena Supplement Catena Verlag, Cremlingen. Vol. 1, 1982.

Natural Resources and Development A biannual collection of recent German contributions concerning the exploration and exploitation of natural resources. Edited by the Institute for Scientific Co-operation in conjunction with the Geological Survey of the FRG and numerous members of German universities, Tubingen. Vol. 1, 1975.

Applied Sciences and Development A biannual collection of recent German contributions concerning development through applied sciences. Edited by the Institute for Scientific Co-operation, Tubingen. Vol. 1 (1973) – Vol. 14 (1979). Continued as *Applied Geography and Development* Vol. 15 (1980).

Zeitschrift für Geomorphologie, N(eue) F(olge) Journal of Geomorphology, New Series.

Important German Series of Handbooks and Textbooks in the Field of Systematic and Regional Geography

Teubner Studienbucher Geographie, Verlag F. Teubner, Stuttgart, 15 vols.

Wissenschaftliche Landerkunden, Wissenschaftliche Verlagsgesellschaft, Darmstadt, 59 vols.

Lehrbuch der Allgemeinen Geographie (founded and) ed. by E. Obst, Verlag de Gruyter, Berlin, 10 vols.

UTB — Universistats-Taschenbucher (Series of paperback-textbooks on the complete spectrum of sciences) appr. 15 vols. of geographical interest.

Das geographische Seminar, Westermann, Braunschweig, 25 vols.

Ertrage der Forschung (i.e. yields of research), Wiss. Verlagsgesellschaft, Darmstadt, appr. 36 vols. of geographical interest.

Wege der Forschung (i.e. ways of research), Wiss. Verlagsgesellschaft, Darmstadt, appr. 32 vols. of geographical interest.

References

Ahnert, F. (1978) 'Gegenwärtige Forschungstendenzen der physischen Geographie', *Die Erde, 109*, 49-80.

Bähr, J. and Mertins, G. (1981) 'Idealschema der sozialräumlichen Differenzierung lateinamerikanischer Grosstädte', *Geographische Zeitschrift, 69*, 1, 1-33.

Bahrenberg, G. and Streit, U. (eds.) (1981) 'German Quantitative Geography', Papers presented at the 2nd European Conference on 'Theoretical and Quantitative Geography' in Cambridge 1980, *Münster'sche Geographische Arbeiten*, Paderborn, 11.

—— Fischer, M.M. and Nijkamp, P. (eds.) (1983) *Recent development in spatial analysis: Methodology, measurement, models*, Gower Publishing Co., Aldershot, Hampshire.

Barsch, D. and Liedtke, H. (1980) 'Principles, scientific value and practical applicability of the geomorphological map of the Federal Republic of Germany at the scale of 1:25,000 (GMK 25) and 1 100,000 (GMK 100)', *Zeitschrift für Geomorphologie, NF., Suppl. 36*, 296-313.

Bartels, D. (1968) 'Türkische Gastarbeiter aus der Region Izmir', *Erdkunde, 22*, 313-24.

—— (ed.) (1970) *Wirtschafs- und Sozialgeographie*, Neue wissenschaftliche Bibliothek, Vol. 35, Kiepenheuer and Witsch, Köln and Berlin.

Blotevogel, H.H. and Heineberg, H. (1976-1981), *Bibliographie zum Geographiestudium*, 4 vols., Schöningh, Paderborn.

Bobek, H. (1962) 'The Main Stages in Socio-Economic Evolution from a Geographical Point of View' in P. L. Wagner and M.W. Mikesell (eds.) *Readings in Cultural Geography*, University Press, Chicago, 218-47.

—— and Fesl, M. (1978) *Das System der Zentralen Orte Österreichs. Eine empirische Untersuchung*, Schriften der Kommission für Raumforschung der Österreichischen Akademie der Wissenschaften, 3 vols., Hermann Böhlaus Nachf., Wien-Köln.

—— and —— (1983) *Zentrale Orte Österreichs 2*, Schriften der Kommission für Raumforschung der Österreichischen Akademie der Wissenschaften, Wien.

Borsdorf, A. (1978) 'Population growth and urbanization in Latin America. Some Comments on Demographic Development and Urban Structural Change', *GeoJournal, 2*, 1, 47-60.

—— and Wilhelmy, H. (1984) *Die Städte Sudamerikas*, Urbanisierung der Erde, 3, Borntraeger, Stuttgart, in print.

Bremer, Hanna (1983) 'Twenty-one years of German Geomorphology', *Earth Surface Processes*.

Breu, J. (ed.) (1970 ff.) *Atlas of the Danubian countries*, Österreichisches Ost- und Sudosteuropa- Institut, Wien.

Bronger, D. (1977) 'Methodological Problems of Empirical Developing Country-Research', *GeoJournal, 1*, 6, 49-64.

Dettmann, K. (1970) 'Zur Variationsbreite der Stadt in der Islamisch-orientalischen Welt. Die Verhältnisse in der Levante sowie im Nordwesten des indischen Kontinents', *Geographische Zeitschrift, 58*, 95-123.

Deutsche Forschungsgemeinschaft (DFG) (ed.) (1978) 'Recent German Research on the Middle East', *The Tübingen Atlas of the Middle East*, Tübingen.

Ehlers, E. (1982) 'Man and the Environment. Problems in Rural Iran', *Applied Geography and Development, 19*, 108-25.

Eriksen, W. (1975) *Probleme der Stadt- und Geländeklimatologie*, Erträge der Forschung, 35, Wissenschaftliche Buchgesellschaft, Darmstadt.

Fliri, F. (1975) *Das Klima der Alpen im Raum von Tirol*, Monographien zur Landeskunde Tirols, Verlag Wagner, Innsbruck.

Freitag, U., Kaiser, K., Manshard, W., Mensching, H., Schätzl, L., Schultze, J. (eds.) (forthcoming) *Afrika-Kartenwerk*, DFG-Project, Borntraeger, Berlin-Stuttgart.

Geipel, R. (1978) *'Curriculum-Research'* in R. Geipel *et al., Das raumwissenschaftliche Curriculum-Forschungsprojekt. Erfahrungen und Ergebnisse der Entwicklungsphase 1973-1976*, Westermann,

Munich.

Geipel, R. and Pohl, J. (1983) 'Umweltqualität im Münchener Norden', Wahrnehmungs- und Bewertungsstudien, *Münchner Geographische Hefte, 49*, 200 pp.

Geography in Switzerland (1980) *La Géographie en Suisse*, ed. by Geographische Kommission der Schweizerischen Naturforschenden Gesellschaft und der Geographisch-Ethnographischen Gesellschaft Zürich, *Sonderband* der *Geographia Helvetica*, 5, Zürich.

Gierloff-Emden, H.-G. (1977) *Orbital remote sensing of coastal and off-shore environments. A Manual of interpretation*, de Gruyter, Berlin-New York.

Giese, E. (1981) 'The development and present stage of research into "quantitative geography" in the German-speaking countries' in R.J. Bennett (ed.), *European Progress in Spatial Analysis*, Pion, London, 91-111.

Grimm, F. (1979) *The Settlement System of the German Democratic Republic*, Report to the Commission on National Settlement Systems of the IGU, Institut für Geographie und Geoökologie der Akademie der Wissenschaften der DDR, Leipzig.

Haase, G. (1977) 'Ziele und Aufgaben der geographischen Landschafts-forschung in der DDR', *Geographische Berichte, 82*, 1-19.

Hagedorn, H. and Thomas, M. (eds.) (1980) 'Perspectives in geomorph-ology', Papers from the First British-German Symposium on Geomorphology, University of Würzburg, 24-29 September 1979, *Zeit-schrift für Geomorphologie, Suppl., 36*.

Hajdu, J.G. (1968) 'Towards a Definition of Post-War German Social Geography', *Annals of the Association of American Geographers, 58*, 397-410.

Hard, G. (1976) 'Physical Geography – its function and future. A reconsideration', *Tijdschrift voor Econ. en Soc. Geografie, 67, 6*, 358-86.

Harris, C.D. (1980) *International List of Geographical Serials* (3rd ed.) Department of Geography, University of Chicago, Research Paper 193, Chicago.

Hassinger, H. (1938) 'Anthropogeographie' in Klute (ed.), *Handbuch der geographischen Wissenschaft*, vol. II, 167-558.

Hellen, J.A. (1978) 'The Future of German Geography', *The Geographical Journal, 144*, 118-21.

Hermanns, H., Lienau, C. and Weber, P. (1979) *Arbeiterwanderungen zwischen den mittel- und westeuropäischen Industrieländern. Eine annotierte Auswahlbibliographie unter geographischem Aspekt,*

Saur, München-New York-London.

Hofmeister, B. (1971) *Stadt und Kulturraum Angloamerika*, Vieweg-Verlag, Braunschweig.

Hottes, K. and Hamilton, Ian (eds.) (1980) 'Case Studies in Industrial Geography', *Bochumer Geographische Arbeiten, 39*.

Hövermann, J. and Wang Ying (1982) 'First Sino-German Joint Expedition to Quinghai-Xizang (Tibet) — Plateau in 1981', *Sitzungsberichte und Mitteilungen der Braunschweigischen Wissenschaftlichen Gesellschaft*, Sonderheft 6.

Hutteroth, W.-D. and Abdulfattah, K. (1977) *Historical geography of Palestine, Transjourdan and Southern Syria in the late 16th century*, Erlangen.

Jäger, H. (1972) 'Historical Geography in Germany, Austria and Switzerland' in A.R.H. Baker (ed.), *Progress in Historical Geography*, C.U.P., Newton Abbot, 45-62 and 218-25.

—— (1981) 'Late Medieval Agrarian Crises and Deserted Settlements in Central Europe', in N. Skynm-Nielsen and N. Lund (eds.), *Danish Medieval History*, New Currents, Copenhagen, 223-37.

—— (1982) 'Reconstructing Old Prussian Landscapes, with special reference to spatial organization', in A.R.H. Baker and M. Billinge (eds.), *Period and Place — Research Methods in Historical Geography*, Cambridge, 33-50, 321-2.

James, P.E. (1972) *All Possible Worlds: A History of Geographical Ideas*, Bobbs-Merrill Company, Indianapolis.

Jentsch, Ch. (1973) 'Das Nomadentum in Afghanistan', *Afghanische Studien, 9*.

Jusatz, H.J. (ed.) (1967-80) *Medizinische Länderkunde — Geomedical Monograph Series*, Vol. 1: Libya, Vol. 2: Afghanistan, Vol. 3: Ethiopia, Vol. 4: Kuwait, Vol. 5: Kenya, Vol. 6: Korea, Springer-Verlag, Berlin-Heidelberg-New York.

Karger, A. and Sperling, W. (1980) 'Die Entwicklung der geographischen Osteuropaforschung', *Osteuropa, 30*, 8/9, 747-52.

Keller, R. (1961) *Gewässer und Wasserhaushalt des Festlandes. Eine Einführung in die Hydrogeographie*, (translated into Russian, Moscow 1965).

—— (1974) 'Man's influence on the Hydrological Cycle of the Federal Republic of Germany', *Man and Environment*, Studies in Geography in Hungary, 11, 149-56.

—— (ed.) (1979) 'Studien zur Landeskunde Tirols und angrenzender Gebiete', Festschrift des Instituts für Landeskunde zum 60. Geburtstag von A. Leidlmair, *Innsbrucker Geographische Studien*, 6.

Kinzl, H. (1940) 'Los glaciares de la Cordillera Blanca', *Revista de Ciencia Natural*, Lima, *43*, 417-40.

Kohlhepp, G. (1981) 'Analysis of state and private regional development projects in the Brazilian Amazon Basin', *Applied Geography and Development*, *16*, 53-79.

Kob, A. (1963) *Ostasien. China-Japan-Korea. Geographie eines Kulturerdteils*, Quelle u. Meyer-Verlag, Heidelberg.

Kopp, H. (1981) *Agrargeographie der Arabischen Republik Jemen*, Erlangen.

Krebs, G. (1982) 'Regional Inequalities during the Process of National Economic Development: A Critical Approach', *Geoforum*, *13*, 71-81.

Krönert, R. (1981) 'City-hinterland regions of large cities and medium-sized towns in the GDR', *Geographia Polonica*, *44*, 139-50.

Krüger, K. (1982) 'Regional Policy in Malaysia', *Geoforum*, *13*, 133-49.

Lauer, W. (ed.) (1976) *El Proyecto Mexico de la Fundacion Alemana para la Investigacion cientifica – bibliografia (1964-1976)*, Wiesbaden.

Lauer, W. (1979) 'Puebla-Tlaxcala. A German-Mexican Research Project', *GeoJournal*, *3*, 1, 97-105.

Lautensach, H. (1945) *Korea, Eine Landeskunde auf Grund eigener Reisen und der Literatur*, Leipzig.

Leitner, H. (1982) 'Residential segregation, socioeconomic integration and behavioural assimilation: the case of Yugoslav migrant workers in Vienna', in European Science Foundation (ed.), *Cultural Identity and Structural Marginalization*, Human Migration II, Strasbourg, 59-78.

Leser, H. (1980) 'Soil erosion measurements on arable land in North-West Switzerland' in *Geography in Switzerland – Collection of Papers Offered to the 24th International Geographical Congress Tokyo August 1980*, Bern-Zurich, 9-14.

Lichtenberger, E. (1975) 'Forschungsrichtungen der Geographie. Das Österreichische Beispiel 1945-1975, Geographie, Kartographie, Raumordnung', *Mitteilungen der Österreichischen Geographischen Gesellschaft*, *117*, 1, 2, 1-116.

—— (1978) 'Quantitative Geography in the German-Speaking Countries', *Tijdschrift voor Economische en Sociale Geografie*, *69*, 362-73.

—— (1979) 'The impact of political systems upon geography: the case of the Federal Republic of Germany and the German Democratic Republic', *Professional Geographer*, *31*, 2, 201-11.

—— (1981) 'The impact of institutional forces on the state of university geography in the Federal Republic of Germany in comparison with Britain', in R.J. Bennett (ed.) *European Progress in Spatial Analysis*, Pion, London, 112-30.

—— (1983) 'Perspectives of urban geography', International Symposium for Research into 'Stadtgestalt', Trier 1979, *Trierer Geographische Studien.*

—— forthcoming *Gastarbeiter – Leben in zwei Gesellschaften*, Ergebnisse von zwei bilateralen Forschungsprojekten über jugoslawische Gastarbeiter in Wien 1974 und 1981, Wien.

Lüdemann, H. (1980) 'Geographische Forschung in der DDR – Entwicklung und Perspektiven', *Petermanns Geographische Mitteilungen*, *124*, 2, 97-103.

Manshard, W. (1974) *Tropical Agricultre, A Geographical Introduction and Appraisal*, Longman Group, London.

—— (1976) 'Recent Trends in German Geography', *Geoforum, 7*, 383-5.

—— 1977) *Die Städte des tropischen Afrika*, Urbanisierung der Erde, 1, Borntraeger, Stuttgart.

Mensching, H. (1974) 'Landforms as a dynamic expression of climatic factors in the Sahara and Sahel – a critical discussion', *Zeitschrift für Geomorphologie, Supple., 20*, 168-77.

—— and Ibrahim, F. (1977) 'The Problem of Desertification in and around Arid Lands', *Applied Sciences and Development, 10*, 7-43.

—— (1982) 'Applied Geomorphology: Examples of Work in the Tropics and the Subtropics', *Applied Geography and Development*, 19, 87-96.

Messerli, B. *et al.*, (1978) 'Fluctuations of climate and glaciers in the Bernese Oberland, Switzerland, and their geoecological significance, 1600 to 1975', *Arctic and Alpine Research, 10*.

Meynen, E. (ed.) (1982) *Sixlingual Synopsis of the 2400 Keyterms.* Submitted at the occasion of the 24th International Geographical Congress Tokyo, 2 vols., F. Steiner-Verlag, Wiesbaden.

Mikesell, M.W. (ed.) (1973) 'Geographers Abroad', Essays on the Problems and Prospects of Research in Foreign Areas, The University of Chicago, Department of Geography, Research Paper, 152.

Mohs, G. (1980) *Migration and Settlement 4: German Democratic Republic*, International Institute for Applied Systems Analysis, Laxenburg/Austria, RR-80-6.

Müller, P. (1974) *Aspects in Zoogeography*, The Hague

—— (1983) 'Unterentwicklung durch "Rentenkapitalismus"'? Geschichte, Analyse und Kritik eines sozialgeographischen Begriffes und seiner Rezeption', *Urbs et Regio, 29.*

Neef, E. (1967) *Die theoretischen Grundlagen der Landschaftslehre,* VEB Haack, Gotha.

Nickel, H.J. (1978) *Soziale Morphologie der mexikanischen Hacienda,* Das Mexico-Projekt der Deutschen Forschungsgemeinschaft, XIV, Wiesbaden (translation into Spanish forthcoming).

Nitz, H.J. (1974) *Historisch-genetische Siedlungsforschung, Genese und Typen ländlicher Siedlungen und Flurformen,* Wissenschaftliche Buchgesellschaft, Darmstadt.

Otremba, E. (ed.) (1962 ff.), *Atlas der deutschen Agrarlandschaft,* Wiesbaden.

Penck, A. (1925) 'Das Hauptproblem der physischen Anthropogeographie', *Zeitschrift für Geopolitick, 2,* 5, 330-48.

Rathjens, C. (1979) *Die Formung der Erdoberfläche unter dem Einfluss des Menschen. Grundzüge der anthropogenetischen Geomorphologie,* Teubner Taschenbücher der Geographie, Stuttgart.

Richter, G. and Negendank, J. (1977) 'Soil erosion processes and their measurement in the German area of the Moselle River', *Earth Surface Processes, 2,* 261-78.

Ruppert, K. (ed.) (1981), *W-G-I-Berichte zur Regionalforschung: 4,* Tätigkeitsbericht 1976-1980 des Instituts für Wirtschaftsgeographie der Universität München.

——(1983) 'The concept of social geography', *GeoJournal VII.*

Sandner, G. (1982) 'La planificación regional integrada como agente del estado frente a la comunidad local y lá patria chica. Un resumen de experiencias centroamericanas', *Seminar on Regional Development Alternatives in the Third World,* Belo Horizonte.

Sanke, H. (1980) 'Zur bicherigen, gegenwärtigen und zukünftigen Entwicklung der regional- und auslandsgeographischen Forschung und Lehre — am Beispiel der Sektion Geographie der Humboldt-Universität', *Zeitschrift der Humboldt-Universität zu Berlin, Math.-Nat. Reihe XXIX/2,* 135-43.

Schätzl, L. (1981) *Wirtschaftsgeographie,* Vol. 1: *Theorie,* Vol. 2:- *Empirie,* Uni-Taschenbücher, 782 und 1052.

Schmidt-Renner, G. (1970) 'Elementare Theorie der ökonomischen Geographie: nebst Aufriss der historischen ökonomischen Geographie: ein Leitfaden für Lehrer und Studierende', Translated into Japanese, Kohm-Shom-Co. Tokyo.

—— *et al.* (1973) *Ekonomičeskaja geografija Germanskaja Demo-*

kratičeskaja Republiki: geografija naselenija i otraslej narodnogo chozjajastva, Izd. mysl., Moskva.

Schneider, S. (1974) *Luftbild und Luftbildinterpretation*, de Gruyter, Berlin.

Schöller, P. (1973) 'Trends in Urban Geography in the German Language Area 1952-1970' in P. Schöller (ed.) *Trends in Urban Geography* (Reports on Research in Major Language Areas), Bochumer Geogr. Arbeiten, 16, 31-41.

—— (1980) 'Centrality and Urban Life in Japan', *GeoJournal, 4, 3*, 199-204.

Scholz, F. (1981) 'Theoretical Remarks on the reason and result of the modern changes in Bedouin/Nomad Societies of the countries in the islamic Orient', Paper prepared for the International Union of Anthropological and Ethnological Sciences Intercongress, Amsterdam.

Schweizer, G. (1974) 'The Aras-Moghan Development Project in Northwest Iran and Problems of Nomad Settlement', *Applied Sciences and Development, 4*, 134-48 and *Zeitschrift für auslandische Landwirtschaft*, 12/1973, 60-75.

Schwind, M. (1967, 1981) *Das Japanische Inselreich: Wirtschaftsgrossmacht auf engen Raum*, Vol. 1: *Die Naturlandschaft*, Vol. 2: *Kulturlandschaft*. De Gruyter, Berlin-New York.

Semmel, A. (1977) *Grundzüge der Bodengeographie*, Teubner-Studienbucher Geographie, Stuttgart.

Stäblein, G. (1980) 'Studies in the periglacial environment: A review of geomorphodynamic, cryopedological and quaternary research in Germany', *Zeitschrift für Geomorphologie, NF., 36*, 84-95.

- —— (1983) 'Geomorphic altitudinal zonation in the arctic-alpine-mountains of Greenland', *Mountains research and development, 3*, Boulder, Colorado/USA, in print.

Stadelbauer, J. (1983) 'Wege, Möglichkeiten und Grenzen der BRD-Regionalforschung in sozialistischen Ländern: Bestandsaufnahme der 70er Jahre und Ausblick', Freiburg-im-Breisgau, manuscript not in print.

Sweeting, M. and Pfeffer, K.-H. (eds.) (1976) 'Karst Processes', *Zeitschrift für Geomorphologie, Suppl., 26*.

Thomale, E. (1972) 'Sozialgeographie. Eine disziplingeschichtliche Untersuchung zur Entwicklung der Anthropogeographie', *Marburger Geographische Schriften, 53*.

Tichy, F. (1979) 'Genetische Analyse eines Altsiedellandes im Hochland von Mexico – Das Becken von Puebla-Tlaxcala' in H-J. Nitz and

J. Hagedorn (eds.) *Gefügemuster der Erdoberfläche – Festschrift zum 42. Geographentag Göttingen 1979*, 339-74.

Troll, C. (1949) 'Geographical Science in Germany During the Period 1933-1945: A Critique and Justification', *Annals of the Association of American Geographers, 39*, 100-37.

Uhlig, H. (ed.) (1983) *Recent Pioneer Settlement – Planned versus spontaneous colonization – in South East Asia;* Part I, Uhlig, H. *Planned and spontaneous settlement in the Asean States of Sea,* Part II, Riethmüller, R., Scholz, U., Sirimbsamband, N. and Spaeth, A. *Two case-studies from SE-Thailand,* Institut für Asienkunde, Hamburg and *Geissener Geographische Schriften.*

—— and Lienau, C. (eds.) (1974) *Basic materials for the terminology of the agricultural landscape*, Deutsch-English-Français, 3 vols., Giessen.

Van Valkenberg, S. (1951) 'The German School of Geography' in Griffith Taylor (ed.), *Geography in the Twentieth Century*, Philosophical Library, New York, 91-115.

Wirth, E. (1975) 'Die orientalische Stadt. Ein Uberblick aufgrund jungerer Forschungen zur materiellen Kultur', *Saeculum, 26*, 1, 45-94.

—— (1974-5) 'Zum Problem des Bazars (suq carsi)', *Der Islam, 51*, 203-60; *52*, 6-46.

—— (1983) 'Geographische Feldforschungen im Orient', *Forschungen in der Bundesrepublik Deutschland. Beispleile, Kritik, Vorschlage*, Deutsche Forschungsgemeinschaft, Weinheim.

Wissmann, H. von (1968) 'Zur Archäologie und antiken Geographie von Südarabien: Hadramaut, Qataban und das Adengebiet in der Antike', *Nederl. Histor.-Archeolog. Instituut in het Nabijc Oosten, 24.*

Zimm, A. (1982) 'Darstellung regionaler Strukturen und Prozesse in der Geographie des sozialistischen Auslands', *Petermanns Geographische Mitteilungen, 126, 4*, 217-22.

Zötl, J. (1974) *Karsthydrolgeologie*, Springer-Verlag, Wien.

9 NORTH AMERICA

Marvin W. Mikesell

When professors and students began or resumed their studies after the end of the Second World War, geography in North America displayed the evidence of national and international influence that is characteristic of the discipline as a whole. For some members of the American geographical profession their calling could be seen as a home-grown enterprise, a product of American minds stimulated by the vast and varied American environment (Colby, 1936). For others the scholarly tradition followed or adopted had deeper and more extensive roots and could be regarded as a product of European as well as American influence. For example, in the most successful textbook of the pre-war decade, James (1935) used the bio-climatic realms of Siegfried Passarge's *Landschaftsgürtel der Erde* (Breslau, 1923) as an organising theme and treated problems of human-environmental relationship within the framework of *possibilisme* that had been erected by French geographers and historians. Similarly, Sauer's attempt in 1925 to redefine human geography as an enterprise devoted to study of how 'natural landscapes' evolve into 'cultural landscapes' was an elaboration of ideas expressed previously by German geographers. Moreover, in at least one American department (Berkeley: Leighly, 1979) Walther Penck's *Die morphologische Analyse* (Stuttgart, 1924) was regarded as a more convincing foundation for geomorphology than the writings of the doyen of the Association of American Geographers (Davis, 1909). In addition, it is probably safe to state that the aspect of geographical lore that was best known among American undergraduates immediately after the war was the climatic classification presented in Wladimir Köppen's *Grundriss der Klimakunde* (Berlin, 1931). Finally, and perhaps most significantly, Hartshorne's *The Nature of Geography* (1939), which had served as a 'declaration of independence' for American geographers, entailed adoption of Alfred Hettner's vision of the place of geography among the sciences.

In spite of these strong European influences, American geography in 1945 displayed evidence of intellectual currents, opportunities and leadership that were peculiar to the northern half of the New World. In subsequent years, the discipline was marked by rapid growth, prolif-

eration of subfields, competition among more-or-less distinctive departments, and eventually by a 'new geography' intended to supplant orientations thought to be deficient in scientific status or explanatory
power. I have tried elsewhere (Mikesell, 1981) to describe the forces
responsible for change in American geography. In the pages that follow
my effort is more narrative in character, an attempt to identify and
depict some of the highlights of North American geography during
the years from 1945 until the early 1980s. Emphasis is placed on major
events that captured the spirit of these times. Like the account of any
participant observer, my effort is more impressionistic than realistic
and it has both the advantage and disadvantage of hindsight.

Restoration and Expansion of the Discipline, 1945-50

In 1945 programmes of graduate training in geography were available
in 28 universities in the United States and also at the University of
Toronto in Canada. By 1947 several other institutions had added graduate programmes, including McGill University and the University of
British Columbia. Enrolment wherever geography was offered increased rapidly. Veterans' benefits, which included subsidies for tuition
and maintenance, permitted many students who might otherwise have
been denied such opportunity to embark upon programmes of graduate
study. Students whose work had been interrupted by the war resumed
their studies and joined expanding faculties after completion of their
dissertations. Exposure through military service to foreign areas and
the post-war spirit of internationalism encouraged enthusiasm for geography that was less evident in the isolationist atmosphere of the pre-
war years.
   Professors facing the challenge of rebuilding or expanding their programmes had both assets and liabilities. Assets included the geographical wisdom stored in 35 volumes of the *Annals of the Association of
American Geographers* and, coincidentally, in 35 volumes of the *Geographical Review*. In the most widely used textbook of the immediate
post-war period (Finch and Trewartha, 1936), the accomplishments and
tasks of the discipline were defined in reference to 'elements' of nature
and culture. The former were probably better understood than the
latter at this time, and an adverse reaction to the environmentalism of
the 1920s discouraged efforts to trace causal relationships between
physical and human geography. The intellectual spirit of the time was
cautious and inductive, and honour was accorded most readily for

arduous field-work.

A doctoral candidate presenting himself for a qualifying examination in 1945 would probably have been expected to display mastery of the Köppen classification, explain the Davisian concept of an erosion cycle, describe and delimit one or more of Fenneman's physiographic provinces (1931, 1938), and exhibit understanding of how the acknowledged 'elements' of geography could be mapped in the field. As a prospective Doctor of Philosophy he (not often she at this time) might also have been expected to expound upon the distinction between environmentalism and possibilism. He would also have been expected (unless he were at Berkeley) to comment on his dog-eared and heavily underlined copy of Hartshorne's *Nature of Geography*. He would not have been expected to display sophisticated understanding of any neighbouring field of social science. Indeed, his minor or prior field would most likely have been botany, geology, meteorology or some other natural science.

For professors and students alike, the major liability of the discipline in the immediate post-war period was the absence of textbooks on the newly emerging subfields of human geography. Yet, ironically, this period was also a time when identification with such subfields became pervasive among American geographers. In the 1930s few scholars thought it necessary to place an adjective before 'geography' or 'geographers'. By 1949 most geographers had adopted one or more qualifying designations: physical, political, economic, historical, cultural, urban and so on. Indeed, by this time certification as a qualified member of the profession usually entailed two labels, one topical and the other regional. In rapidly expanding departments, such as the one at Los Angeles, new faculty were being recruited on the basis of their ability to contribute to such hybrid enterprises as the political geography of Eastern Europe, the urban geography of North America and the cultural geography of Latin America.

The literature available to students of urban geography exemplifies the problems and prospects of the time. The only book on this topic was the idiopathic work of Taylor (1949), although Dickinson (1947) had offered suggestions on how city-hinterland relations might be analysed and anticipated later emphasis on the concept of urban hierarchy. Attention was directed primarily to article-length expositions of problems thought to be appropriate for geographical analysis, such as Colby's 'Centrifugal and Centripetal Forces in Urban Geography' (1933) and Harris and Ullman's 'The Nature of Cities' (1945). Ullman's 'A Theory of Location for Cities' (1941), which more clearly anticipated later interest, was less influential among geographers, a consequence perhaps of its publication in an 'alien' journal. Harris's 'A Functional Classification of Cities in the United States' (1943) offered more immediate opportunity for verification and elaboration. But it was probably monographic works, mainly dissertations, that constituted the primary reference material for the urban geography of the 1940s. The first summary statement on the geography of cities, an anthology rather than a textbook, appeared much later (Mayer and Kohn, eds., 1959), when urban geography was already well established as a professional field.

Debate among academic geographers during the 1940s centred primarily on Hartshorne's strictures against confusion of the missions of geography and history, which had provoked a critical response from Sauer (1941). Hartshorne's more important message, a denial of the prospect that geography could be defined in reference to presumed causal relationships between natural environment and human activity, provoked little criticism, although 'environmentalism' could still be found in some textbooks (White and Renner, 1945). More attention was directed to the question of how the Association of American Geographers (AAG), previously regarded as an elite organisation open only to scholars of demonstrable accomplishment, might be transformed into a voluntary association open to anyone interested in or dedicated to the advancement of geography (James and Martin, 1978). The merger in 1948 of the AAG and ASPG (American Society for Professional Geographers) resolved that controversy. By the end of this frenetic decade, American geographers could settle down to the task they had set for themselves: to offer a disciplined interpretation of 'the areal differentiation of the world' (Hartshorne, 1939, p. xi).

Inventory and Prospect, the 1950s

The 1950s was a time of rapid, almost explosive, growth. Membership in the AAG increased from 1,300 in 1949 to 1,838 in 1957 and passed 2,700 in 1963. The dominant position of the departments at Berkeley, Chicago, Clark, Michigan and Wisconsin was challenged by new or expanded graduate programmes at UCLA, Georgia, Louisiana State, Minnesota, Nebraska, North Carolina, Ohio State, Pennsylvania State, Syracuse and Washington. In Canada the opportunities for graduate work already available at Toronto, McGill and British Columbia were enhanced by programmes established at Alberta, Calgary, Manitoba,

McMaster, Ottawa and Western Ontario. In 1951 the quality of Canadian geography was also enhanced by the establishment of the Canadian Association of Geographers.

By 1952 the literature produced by scholars working in each of the dozen or so well-recognised subfields of the discipline had proliferated to such an extent that an assessment of accomplishments and unfinished tasks seemed desirable. The volume produced in response to this need, *American Geography: Inventory and Prospect* (James and Jones, eds., 1954), was and still remains the most comprehensive review undertaken by American geographers. Opportunity for assessment was also presented by the International Geographical Congress held in Washington DC in 1952. Papers offered to that congress and the guidebooks prepared for congress excursions ('Industrial Cities', 'New England', 'Southeastern', Transcontinental') permitted American geographers to display their wares and, for excursion leaders, to offer informed commentary on the richly varied American scene. But it was the 'official' survey published on behalf of the AAG in 1954 that best depicted the spirit and energy of the profession.

The 26 chapters in the volume, some written by single authors and others produced by committees, indicate preoccupation with topical fields: urban geography, agricultural geography, climatology and so on. Regional study (Whittlesey, 1954) is presented in a non-specific overview intended to be a guide for investigation at several scales. The position of this essay, following James's introductory chapter on 'The Field of Geography', suggests that it was regarded by the editors of the volume, and perhaps by the AAG, as the most important or at least the most provocative contribution to the survey. In many respects this non-specific overview was an attempt to define the character of a non-field. Regional study was of course well presented in the lore of the discipline, but few later efforts were made to treat regions generally rather than specifically.

It is also significant that *American Geography: Inventory and Prospect* includes chapters devoted to elements of both physical and human geography but no chapter devoted to their relationship. The once keenly debated issue of environmentalism had so receded from professional consciousness that it was not thought to be deserving of further comment. As James (p. 13) remarked: 'In a relatively stable environment, no example of positive determinism has been demonstrated by acceptable method and consequently the concept is no longer considered useful as a guide to geographic understandings'. In 1954 the spirit of Huntington and Semple no longer haunted members of the AAG, and the neo-environmentalism perhaps already latent in the smog of Los Angeles and the insidious movement of DDT had not yet captured their imagination. Literature was of course available elsewhere for those few geographers who wanted to continue or revive the old debate. Indeed, in a rival survey, Tatham (1951) discussed 'Environmentalism and Possibilism' in sceptical but open-minded terms that would have greatly enhanced the quality of *American Geography: Inventory and Prospect.*

If one takes James's introductory essay as a statement of professional consensus and regards the structure and emphasis of the 1954 AAG survey as a fair assessment of professional commitments, then it is clear that American geographers at this time were devoted primarily to the cultivation of topical fields. They were expected to become expert in the interpretation of urban, rural, commercial, industrial, physical or biological phenomena. They were also expected to be aware of the operation of historical processes that alter the character or significance of these phenomena and to master methods or techniques for disciplined inquiry in the library or field. Their essential obligation was to understand points, lines and areas on maps. The sprawl already evident in geographical inquiry had taken many scholars towards and even into neighbouring fields, but geographers were still admonished to 'adhere to the geographic point of view and practise incontrovertible geographic methods'. The great challenge for the profession was to bring to systematic studies 'the touch of reality that springs from the regional concept'.

Another major survey (Thomas, ed., 1956) provided a different perspective on the interests of American geographers. To be sure, this volume, *Man's Role in Changing the Face of the Earth*, was not intended to be a review of geography or a forum for geographers. Yet geographers are so prominent in the volume that one must see this international and interdisciplinary effort as a major professional accomplishment. Inspired by the pioneer work of Marsh (1864), the contributors to *Man's Role* sought understanding of the processes (deforestation, erosion, reclamation, air and water modifications) that have altered the world's physical and biological environments. In contrast to the efforts of earlier environmentalists, an attempt was made by these authors to assess the cumulative ability of the human species to create artificial environments and the influence of such environments on human well-being. The ability of geographers to respond to this grand theme, both empirically and philosophically, had a salutary effect on professional morale. Yet the publication was out of phase in reference to larger public interest. Had it appeared a dozen years later, on the eve of Earth Day, geographers might have enjoyed the public recognition that they have often yearned for but seldom achieved.

Blessed with the wisdom of hindsight, it is possible to detect many themes in the literature of the 1950s that anticipated later interests. For example, Christaller's work is cited in the chapters devoted to settlement (Kohn, 1954) and urban geography (Mayer, 1954) in *American Geography: Inventory and Prospect*, as is Weber's work in the chapter devoted to industrial location (Harris, 1954). More prophetic, perhaps, was Leighly's assessment (1954) of the importance of the work undertaken in Sweden on diffusion of innovations: 'No one who essays in the future to interpret the distribution of culture elements in process of diffusion can afford to ignore Hägerstrand's methods and conclusions.' Such hints of 'things to come' abound in the literature of the 1950s. In addition to transparent hints, such as Brush's study (1953) of the hierarchy of central places in South-western Wisconsin, it is possible, with added effort, to find antecedents of most of the interests that captivated the 'new geographers' of the coming decade. However, the notion that the novelties of that decade were not really so novel can be pushed too far, for American geographers were not yet embarked on their later headlong rush into social or, still less, behavioural science. Most members of the profession were devoted to tangible, mappable objects and hence were followers, however unwittingly, of the dictum that geography concerns places or patterns rather than people *per se*. American geography in the 1950s was still 'earthbound'. Yet the drift away from traditional interests was already so advanced that one member of the profession (Leighly, 1955) was moved to ask 'What has happened to physical geography?' A decade later this question was often recast in more ominous terms: 'What has happened to geography?'

'Old' and/or 'New', the 1960s

It is tempting to discuss the epic of the 'new geography' under the heading that Brown (1948) used for discussion of the environment of the Great Plains: 'What it was and what it was thought to be.' With the advantage of hindsight and a growing 'revisionist' literature (Taaffe, 1974; Sack, 1974; King, 1976), it seems clear that many of the claims made on behalf of work published in the 1960s were misleading. The often-repeated assertion that this was a time of revolution is especially open to doubt. Since antecedents are so clearly evident in earlier decades, evolution would be a more judicious term. Or perhaps reformation if a sense of drama is desired. Reformation seems especially suitable because it permits use of the equally plausible notion of counter-reformation. In any event, the 1960s was a decade of remarkable turbulence and heated debate. That this turbulence receded and debate ceased to be heated within four or five years suggests that the controversy about 'old' and 'new' geography was both initiated in and terminated during the 1960s.

In spite of real (Bunge, 1979) or imagined (Gould, 1979) difficulties, it seems clear that the advocates of quantification, model building and computer use enjoyed prompt recognition and rapid advancement. The first publication to offer an effective preview of the methods, aspirations and rhetoric of the 'new geography' may have been a brief and highly tentative study of the location of agricultural activities (Garrison and Marble, 1957). By the end of the following year papers by Berry and Garrison (1958a and 1958b) had offered a convincing demonstration of the potentials of central-place theory, and Walter Christaller had become an ubiquitous guru. Although full comprehension of *Die zentralen Orte in Süddeutschland* (Jena, 1933) was delayed until a translation of it was published in 1966, graduate students soon became as well aware of Christaller's hexagons as they had been familiar with Köppen's climatic formulae. Garrison's review of the 'Spatial Structure of the Economy' (1959-60) encouraged an extension of the 'new' inquiry, as did several elaborations of Thünen's venerable model. Other European location economists, most notably August Lösch and Alfred Weber, also achieved the status of '*post mortem* geographers *honoris causa*' (Krumme, 1970, p. 4). This happy designation could not be applied to the American location economist, Walter Isard, whose *Location and Space Economy* (1956) had preempted much of the territory of the 'new geography' and whose growing empire of regional science could be regarded as a threat to the status of geography. This potential embarrassment may have been dispelled when geographers decided to join Isard's movement rather than compete with it. In any case, when Kuhn issued his manifesto in 1962, new geographers found it easy to endorse the flattering notion that they were part of a revolutionary paradigm. A year later it was possible for a sympathetic observer (Burton, 1963) to declare not only that a revolution had taken place but also that it had already been concluded. The same self-confident spirit was expressed more precisely (Ackerman, *et al.*, 1965, p. 4) when 'location theory' was applauded as an approach capable of synthesising older fields of urban, economic and transportation geography. In such studies, the 'dialogue' between the empirical and theoretical was held to have 'gone the farthest of any problem area of geography'.

Why did the 'new geography' of the 1960s enjoy such rapid and pervasive success? And why, after only a few years of effort, were its advocates able to occupy positions of substantial influence and authority? The most immediate thought inspired by these questions is that the 'new' of this time was better than the 'old', that it was more scientific or at least more verifiable than previous empirical studies, was more likely to lead to general rather than particular findings, and hence had the potential to be expressed in cumulative rather than additive literature. Success may also have encouraged the fact that the 'new geography' was only procedurally or technically and not substantively new. The problems addressed — size and spacing of cities, location of commercial establishments, zonation of land use — were issues that had long been of interest to geographers. Moreover, the initial focus of effort, in urban and economic geography, produced heavy impact on two of the most populated fields of American geography (Mayer and Kohn, eds., 1959; McCarty and Lindberg, 1966). Had the 'new geography' been directed toward peripheral or less crowded fields, its impact would undoubtedly have been less substantial. In any event, once established in core areas of professional concern, the 'new geography' proved to be capable of both territorial expansion and personal co-option. The latter capability is illustrated by Harvey's (1967) conversion of Sauer's (1952) speculations on agricultural origins into a flow chart that resembles a Victorian plumbing system.

It is also possible that an event not often associated with this academic adventure played a key role in its success. The suggestion, made earlier, that 1957 might have been the inaugural year of the 'new geography' invites speculation on the significance of another event occurring in that year: the launching from Kazakhstan of the world's first artificial satellite. The American response to that disturbing accomplishment was a massive effort to improve the quality of scientific education. The benefits accruing to geography may have been trivial compared to what was offered to the disciplines of physical science and engineering, yet they were sufficiently generous to permit numerous summer institutes designed to help geographers to master quantitative techniques and to upgrade the scientific quality of their discipline. In this atmosphere of financial largesse and governmental sponsorship, numerous geographers acquired at least rudimentary understanding of statistics and computer technology. The same atmosphere also explains two larger enterprises sponsored by the AAG: the Commission on, College Geography (1953-74) and the High School Geography Project (1961-70). The 1960s was an ideal time not only for scientific innovation, but also for rapid diffusion of such innovation within professional associations. Had the 'new geography' appeared at an earlier time of less rapid professional growth or a later time of less generous subsidy its impact would probably have been muted and its propagation might have been curtailed.

Reference to something 'new' implies awareness of something 'old'. The opposition implied by these terms is misleading. It is more revealing to speak of 'new' and 'continuing' geographies. Of any ten geographers attending an annual meeting of the AAG in the mid-1960s, probably only two or three would have been self-proclaimed 'new geographers'. Another two or three of this hypothetical group would have been cultural or historical geographers aware of, but not captivated by, computers, quantification or model building. Another of the ten would have been a foreign-area specialist, who would have bridled at the thought that his work was 'old-fashioned or merely descriptive', and, unless his interest were in Europe, would also have been prompt to express scepticism about the transferability of Western models and the quality of numerical data. Add one or two physical geographers and any one of the extraordinary array of specialists usually encountered during AAG meetings and the prominence of 'new geographers' would not have been as conspicuous as published statements might have led an outsider to believe.

Hindsight also permits the comment that geography in the 1960s did not advance in the directions that were often predicted at the beginning of the decade. At present probably only three or four geographers in all of North America deserve to be labelled as accomplished 'mathematical geographers'. The prediction, plausible in 1960, that geography might become a branch of applied mathematics did not materialise. Nor had geography become geo-econometrics, social physics, or an enterprise so abstract in its accomplishment that it no longer concerned the real world. Doubt was eventually expressed (Sack, 1974) on whether a spatial orientation had replaced earlier interests that were more specifically areal or focused on particular places. Thanks to the preoccupation with planning problems that soon became evident among both 'new geographers' and regional scientists one can even express doubt on whether geography in the 1960s was more theoretical than empirical or more basic than applied.

In short, American geography emerged from the turbulent decade of the 1960s with continuity more conspicuous than discontinuity. The methods of the 'new geography' had been absorbed and would soon be evident even in introductory textbooks (Abler, Adams and Gould, 1971; Kolars and Nystuen, 1974). The leaders of the movement, in spite of initial anxiety and consequent aggressiveness, found secure places within the profession. As future chairmen, deans, editors and, eventually, even presidents, they were also destined to become members of the American geographical establishment.

Other Developments in the 1960s

Emphasis on the controversy engendered by the presumed opposition of 'old' and 'new' geographies masks a number of important developments in the 1960s. The 'diverse opinions regarding the nature of geography' that had moved Sauer to launch a reform movement in 1925 continued to be evident, and the discipline continued to be marked by proliferation of subfields. Progress was especially rapid in historical geography. In his chapter on 'Historical Geography' in *American Geography: Inventory and Prospect*, Clark (1954) was obliged to comment on a vast array of work that was more commonly regarded as 'cultural geography'. During the 1960s publications by Clark (1968) and his students (Merrens, 1964; Harris, 1968) and by Meinig (1968) provided a broad foundation for a North American school of historical geography. Clark's leadership was also evident in 1971 when monographs by Meinig and Ward inaugurated the 'Historical Geography of North America Series' of Oxford University Press. When the *Journal of Historical Geography* was launched in 1975, with Clark serving as 'Editor for the Americas', historical geographers working in or devoted to North America had ample reason to be confident of the prospects for their subfield. The publication of the proceedings of a conference on research in historical geography, held in Washington DC at the National Archives (Ehrenberg, ed., 1975), was additional evidence of prospective well-being.

Cultural geographers derived similar benefit from the Prentice-Hall Series on the 'Foundations of Cultural Geography', which was initiated in 1967 with Sopher's volume on the geography of religions. Later volumes by Rappaport (1969), Isaac (1970), Wagner (1972), Zelinsky (1973) and Hart (1975) offered a generous sample of the thinking of cultural geographers, as did Spencer's monograph of 1966 on shifting cultivation and Sauer's *Sixteenth Century North America* (1971), published 15 years after his nominal retirement. The anthology edited by Wagner and Mikesell in 1962 was an interim attempt to define the character of cultural geography, and hence was comparable to the earlier effort of Mayer and Kohn (1959) to offer an interim overview of urban geography. Criticism directed toward Wagner and Mikesell's essentially retrospective view of the field, most notably by Brookfield (1964), had salutary influence on cultural geography, as did the appearance of the first textbooks (West and Augelli, 1966; Spencer and Thomas, 1969) written from a cultural-geographic point of view. However, the most emphatic influence on American cultural geographers was produced by the work of Brookfield (1962) and others (Clarke, 1971; Waddell, 1972) in New Guinea, which encouraged rapid development, extending well into the 1970s, of an effort that came to be known as 'cultural ecology' (Mikesell, 1978).

Among other developments significant in American geography in the 1960s, mention must be made of the programme of natural-hazards research that was initiated by Gilbert White (1958) and soon became a substantial enterprise (Mitchell, 1974; White and Haas, 1975). That the scope of this inquiry was not confined to natural hazards, nor even to geography, however defined, is evident in the interdisciplinary anthology assembled by two of White's students (Burton and Kates, eds., 1965). Natural hazards research and, more broadly, resource-management studies were an important complement to, if not an integral part of, the 'new geography' of the 1960s.

This generalisation also applies to interests subsumed under the heading of 'environmental perception'. As Bowden (1980) suggests, antecedents of this movement can be traced to the 1920s. However, perception became conspicuous as a label for geographical work in the 1960s. Lowenthal's review of 1961 and the special session on 'Environmental Perception and Behavior' held during the AAG meetings of 1965 (Lowenthal, ed., 1967) encouraged geographers of diverse prior commitments to focus attention on the distinction between appearance and reality. By the time Saarinen prepared a review of perception literature for the Commission on College Geography (1969), it was clear that American geography had a new field or, perhaps more accurately, a new way of looking at virtually all existing fields. As a convergent movement, occurring at a time when geography as a whole seemed to exhibit alarming divergence, environmental perception served as a welcome platform on which geographers from seemingly disparate backgrounds could express some measure of common interest.

Persistent Pluralism in the 1970s

Reform movements in American geography have usually been led by persons eager to tidy up an untidy discipline. Yet such opinion always gives rise to the counter-argument that variety is the spice of life. Since the 1960s were marked by attempted reform, it is not surprising that the following decade was characterised by an exuberant pluralism, which was evident not only in proliferating new fields but even in well-established old fields (Butzer, 1973). The validity of this generalisation is evident in the programme of any of the AAG meetings held in the 1970s. The pluralism of this decade was also enhanced by an open convention policy that gave each AAG member the right to be heard, regardless of the novelty or antiquity of his views.

The most conspicuous development during the decade was an expansion of interest in environmental studies. As the success of the 'new geography' of the 1960s was enhanced by national concern fostered by the launching of the first sputnik, so the heightening of environmental interest in the 1970s was a consequence of fall out from Earth Day. No one has yet explained why public interest in environmental problems became so acute in 1970. Concern about air quality was evident in Southern California in the 1950s and concern about chemical hazards spread rapidly after the publication in 1962 of Rachel Carson's *Silent Spring*. In any event, it was understandable that American geographers should want to join less sophisticated neighbours in the environmental crusade that was fostered by public clamour in 1970.

The specific AAG response included a reassessment of physical geography (Carter, Schmudde and Sharpe, 1972), a comprehensive overview of *Perspectives on Environment* (Manners and Mikesell, eds., 1974) and, eventually, a massive *Sourcebook on the Environment* (Hammond, Macinko and Fairchild, eds., 1978). In contrast to earlier orientations in American geography, the contributors to these volumes attempted to see elements of nature and culture not as opposing forces or separate entities but rather as interlocking components of ecosystems. As in the case of perception studies, the environmental guidance offered by geographers in the 1970s represented a convergent movement from several parts of the profession (Lowenthal, *et al.*, 1973). Geographers concerned with the effects of human activities on vegetation and soil began to take account of man's impact on air and water, while geographers devoted to perception studies began to investigate man-made as well as natural hazards. Physical geographers also began to take account of current environmental problems and applied old and new skills in studies of modified climates and landforms. Like the person who suddenly discovers that his old clothes have become fashionable, the environmental geographer of the early 1970s was pleased and somewhat startled by unaccustomed approval (Marcus, 1979). When the preparation of 'environmental impact statements' was compelled by Federal legislation, even 'old-fashioned merely-descriptive regional geography' seemed to have renewed applicability. The cumulative effect of these developments was salutary. Yet the opportunity that became available was for study of modified or polluted environments and not for the pure physical geography of earlier years. Fortunately, as the rhetoric of crisis ecology faded from public memory, geographers able to offer professional expertise on persistent environmental problems could still profit from the national concern that had been fostered by the decidedly unprofessional atmosphere of Earth Day (Detwyler and Marcus, 1972; Terjung, 1976; Gersmehl, 1976; Butzer, 1980).

In addition to expansion of interest in environmental studies, the pluralism of the 1970s was also expressed by accomplishments in regional geography. The 22nd International Geographical Congress, held in Montreal in 1972, encouraged geographers in both Canada and the United States to offer informed commentary on their respective countries. The Canadian effort, facilitated by a prior work issued on the occasion of the Centennial of the Canadian Confederation (Warkentin, ed., 1968), was an impressive series of monographs on 'The Atlantic Provinces', 'Québec', 'Ontario', 'The Prairie Provinces', 'British Columbia' and 'The North' (Trotier, ed., 1972).

The gift offered to members of the Montreal Congress from the USA was a special issue of the *Annals*, also published as a book (Hart, ed., 1972), which consisted of well-documented articles on each of the major regions of the country, plus additional essays on the United States' small towns and metropolitan regions. Interest in the Hispanic realm of North America received a comparable stimulus when the Conference of Latin Americanist Geographers held its initial meeting in 1970 (Lentnek, Carmin and Martinson, eds., 1971). While it cannot be said that these publications restored regional geography to the high status it once enjoyed, they undoubtedly displayed the eclectic virtues of such study in an unusually attractive light.

Of the many additional headings that might be used to describe the pluralism of the 1970s, two deserve special attention: behaviouralism and humanism. Once geographers had accepted the challenge to deal with people as well as mappable objects, they were compelled to move into the once foreign territory of behavioural science. And once they began to conduct interviews, as well as observe and map, they were obliged to take account of the quality of their questions and the problems entailed in interpreting answers. Awareness of the need for sophistication in survey methodology, like the awareness that came early in the development of the 'new geography' of a need for sampling theory, carried geographers into the core area of modern social science. To enquire why people grow what they grow, shop where they shop, or vote for whom they vote is to pose the more significant question of what they think, believe or feel.

During the latter half of the 1960s and throughout the 1970s so many geographers asked questions of this character that thoughts of a 'behavioural revolution' could be entertained. Perhaps the most revealing of the many publications on behavioural geography during the 1970s was the product of a symposium on 'Multidimensional Analysis of Perceptions and Preferences' held during the AAG meeting of 1973 (Golledge and Rushton, eds., 1976). If viewed in reference to the preoccupations of the 'new geographers' of the 1960s, the aspirations of the behavioural geographers of the following decade displayed a significant evolution. Emphasis was placed initially on information and ability to use information as constraints on location decision-making (Pred, 1967-9). In time, behavioural and perceptual geography was expanded to include a wider range of issues, such as multi-dimensional scaling, mental maps, residential preferences, awareness of choice in resource management and both individual and group responses to opportunity or stress.

It seems clear that 'spatial preference' carried a different and perhaps richer meaning than 'spatial structure', as did 'spatial behaviour' in contrast to the once ultra-fashionable concept of 'spatial organisation' (Taaffe, ed., 1970). Yet the challenge implicit in behavioural geography was not fufilled in the 1970s and remains unfulfilled today, in part because training requirements, including serious study of social psychology, differ so notably from traditional conceptions of what should be included in geographical education.

Humanistic geography, like behavioural geography, can be regarded as a predictable, or at least as an understandable, venture. As the behavioural movement implied rejection of the notion, often implicit in geographical writing, that landscapes are created by robots, so the humanistic movement implied rejection of the idea that human behaviour is totally or inherently rational. Love, hate, comfort, fear, patriotism, piety and other comparable *feelings* are usually thought to be beyond the reach of science. Yet they are no less potent for this reason. The debate on whether value-charged issues should be included in human geography did not begin in the 1970s, but it was unquestionably carried to new heights in this decade, perhaps because of adverse reaction to the aggressively rational and deterministic stance of the 'new geography' of the 1960s.

Much of what has been published under the heading of humanistic geography merely calls for greater awareness of human values and individuality. This thought is often expressed not in reference to what people think, see or feel but rather in reference to what a particular author thinks, sees or feels (Samuels, 1979). It is not unfair, consequently, to regard humanistic geography as a movement and perhaps even as a retreat from objectivity towards subjectivity. In any case, Buttimer's *Values in Geography* (1974) and Tuan's *Topophilia* (1974) were soon complemented by statements from a wide range of geographers (e.g. Guelke, 1974; Zelinsky, 1975; Wolpert, 1976; Entrikin, 1976) that called for increased awareness of the extra-scientific or non-positivist dimensions of geographical inquiry. The 'inventory and prospect' so clearly needed for humanistic geography has yet to be accomplished, although two recent anthologies (Ley and Samuels, eds., 1978; Buttimer and Seamon, eds., 1980) can be welcomed as commendable preliminary efforts.

If humanism is extended to include ideology and concern about what can be regarded as justice, then it also includes much of 'radical geography' (Peet, ed., 1977). Although American radical geographers have been notably unsuccessful in their evangelical efforts, they have enjoyed considerable success in demonstrating that there is at least implicit ideological meaning in most human-geographical writing.

The suggestions offered by humanistic and radical geographers have undoubtedly enhanced geographical appreciation. Whether they have enhanced geographical science is open to doubt, for the literature produced so far has been dismissed by most members of the profession as a hedonistic or polemical digression. Yet the epistemological arguments made by these two minority groups have been sustained, for the notion of value-free, totally-objective human geography has almost disappeared from the American academic scene.

Several additional developments in the 1970s demonstrate the diversity and vigour of North American geography. For example, in a review of work in historical geography, Conzen (1980) indicates that 732 article-length publications appeared during the decade. Accomplishment was perhaps even more impressive in urban geography. Urban research figured prominently in the main professional journals, especially in *Economic Geography* and *Geographical Analysis*, and a new journal of *Urban Geography* was launched in 1980. Moreover, the 'Comparative Metropolitan Analysis Project' carried to completion during the decade was the most substantial activity ever undertaken by the AAG. The publications issued included a volume on *Urban Policy-making and Metropolitan Dynamics* (Adams, ed., 1976a), four additional volumes containing 20 urban 'vignettes' (Adams, ed., 1976b) and an atlas (Abler, ed., 1976). By the end of the decade teachers offering urban courses could also profit from the contrasting overviews of Berry (1973) and Harvey (1973) and choose from a wide range of textbooks (e.g. Vance, 1977; King and Golledge, 1978; Hartshorn, 1980; Palm, 1981). Urban research in the 1970s also reflected the heightened sense of social accountability that had been inspired by the civil rights movement and other expressions of public concern during the 1960s. Awareness of defects in the social fabric of American cities, especially in reference to the employment opportunities and residential options of Blacks, was displayed in many publications (e.g. Rose, 1971, 1976).

Progress in smaller or less well-cultivated fields is indicated by the contrast between the speculation about potential uses of remote-sensing technology that was offered in the 1960s and the tangible evidence of such use that was later displayed (e.g. Holz, 1973; Lintz and Simonett, eds., 1976). Also revealing of progress in the 1970s is the contrast between the statement on medical geography offered by May in 1954 and the wide range of both epidemiological and health-care studies later pursued under this heading (Hunter, 1974; Pyle, 1979; Meade, ed., 1980). A connection between remote sensing and medical geography has yet to be made, yet progress in these two detached foci of geographical concern can be taken as examples of specific development during the decade.

An attempt to assess the scope of human geography in the 1970s (Zelinsky, ed., 1978) included consideration of 'Geography as Human Ecology', 'Social Geography', 'Geography and the Mind', 'The Growth of Cultural Geography', 'Western Geography and the Third World', 'The Historical Geography of North American Regions', 'Geography and the Study of Cities' and 'The Spatial Economy'. Although the diversity evident in this set of titles represents a generous sample of professional interest, it only hints at the full range of human-geographical inquiry. A more general assessment, requiring consideration of work in a dozen additional subfields, would still fall short of a complete inventory, for many of the more impressive accomplishments in human geography have reflected individual interest or leadership that is difficult to relate to recognized subfields (e.g. Glacken, 1967; Wheatley, 1971; Schwartzberg, 1978; Jordan, 1981). Moreover, emphasis on progress that is easy to categorise minimises the importance of hybrid enquiries, such as urban biogeography (Schmid, 1975) or historical-political geography (Knight, 1977), and also masks the disagreement that is often evident among scholars working in well-cultivated fields.

During the special programme offered in celebration of the 75th anniversary of the founding of the AAG, statements were made on behalf of the past (James and Martin, 1979), present (Mikesell, 1979) and future (Adams, 1979) of the discipline. It is perhaps significant that the first of these presentations concluded with a reference to creative tension, while the second stressed the difficulty of seeing unity in the diversity of geography, and the third emphasised four categories of potential conflict: between learning and technique, between the profession and its clients, between old and new commitments and between the self-centredness of local-area study and the world-centredness of foreign-area study. Evidence that North American geographers were willing to address these categories of tension, difficulty or conflict was promptly revealed, and from both sides of the border, when publications devoted to *Rethinking Geographical Inquiry* (Wood, ed., 1981) and *Order and Skepticism* (Szymanski and Agnew, 1981) were issued by York University and the AAG.

Limited Growth and New Opportunities Since 1980

Membership in the AAG reached a peak of 6,994 in December of 1975, a reflection of steady growth during the post-war period. Thereafter, it began to decline, mainly because of loss of (graduate) student members. The membership in December of 1982 was 6,010, which was close to the figure for December of 1972. The curtailment of growth experienced by the AAG in the latter half of the 1970s reflected several adverse trends, including curtailment of academic employment opportunities. During the period from 1945 to 1975 programmes of graduate instruction in geography were initiated in 118 North American colleges and universities. After 1975 only 7 new programmes were initiated in the United States and none was initiated in Canada. The post-war period of rapid growth, sustained by the steady increase of the number of Americans of college age, had ended. In an atmosphere of accelerating inflation and curtailment of financial support, American geography was soon destined to be poorer as well as smaller.

The response of the American geographical profession to these developments was to look outward, to search for non-academic applications of geographical findings and skills. In short, the atmosphere of curtailed growth, evident by the mid-1970s, fostered intense discussion about and rapid development of 'applied geography'. An initial conference of applied geographers held at Binghamton in 1978 was repeated in the following year and then became a well-organised annual event. By 1979 geographers engaged in non-academic work (e.g. environmental planning, business-location studies) were as numerous as professors in the applied geography conferences. The profile of AAG membership in 1981 indicates that college or university teaching was still the most important occupation, i.e. 44 per cent of the total membership. If student membership is added to this figure, the combined academic total was a more commanding 66 per cent. But other categories reflected the growing importance of non-academic employment. In that year, 11 per cent of the AAG members were employed by federal and other government agencies, and 8 per cent were in private industry. The hope, now widely shared, that the latter figures might increase substantially may not be realised. In any event, interest in what applied geographers have done or might do has already attracted considerable attention (Myers, 1976; Frazier, ed., 1982).

Although applied geography is the most important focus of current attention, it is not likely to supplant academic interests or inhibit long-standing research commitments. For example, three of the more impressive publications appearing early in the 1980s are expressions of prolonged effort (Sopher, ed., 1980; Brown, 1981; Rooney, Zelinsky and Lauder, eds., 1982). In contrast, several 'Resource Publications' issued recently by the AAG (Morrill, 1981; Calzonetti and Eckert, 1981; Holcomb and Beauregard, 1981; Vale, 1982; Furuseth and Pierce, 1982) have the potential to urge American geographers into new undertakings. Even geopolitics, once thought to be extinct, has enjoyed a respectable rebirth among French-Canadian geographers (Beauregard, ed., 1980).

That the pluralism of the 1970s is likely to continue is suggested by the contrasting tone of recent AAG presidential addesses. Berry (1980) urged his audience to look beyond their discipline and ahead of their time, whereas Hart (1982) offered a plea for restoration of scholarly virtues that may have been evident within the profession in an earlier time, and Helburn (1982) spoke of issues important to the quality of life that demand better scholar-citizenship. There is every reason to believe that the combination of persistence and flexibility, always evident in North American geography, will continue to be revealed in the 1980s.

References

Abler, R.F. (ed.) (1976) *A Comparative Atlas of America's Great Cities*, University of Minnesota Press, Minneapolis.
—— Adams, J.S. and Gould, P.R. (1971) *Spatial Organization: The Geographer's View of the World*, Prentice-Hall, Englewood Cliffs, N.J.
Ackerman, E.A. *et al.* (1965) *The Science of Geography*, National Academy of Science and National Research Council, Washington DC.
Adams, J.S. (ed.) (1976a) *Urban Policymaking and Metropolitan Dynamics: A Comparative Geographical Analysis*, Ballinger, Cambridge, Mass.
—— (ed.) (1976b) *Contemporary Metropolitan America: Twenty Geographical Vignettes*, Ballinger, Cambridge, Mass.
—— (1979) 'The A.A.G. at 75: Future Prospects', *Professional Geographer, 31*, 360-3.
Beauregard, L. (ed.) (1980) *La Problématique géopolitique du Québec*, Cahiers de Géographie du Québec, numéro spécial, Presses de l'Université, Laval, Québec.
Berry, B.J.L. (1973) *The Human Consequences of Urbanization*, St. Martins, New York.
—— (1980) 'Creating Future Geographies', *Annals of the Association of American Geographers, 70*, 449-58.
—— and Garrison, W.L. (1958a) 'The Functional Bases of the Central Place Hierarchy', *Economic Geography, 34*, 145-54.
—— and —— (1958b) 'Alternate Explanations of Urban Rank-Size Relationships', *Annals of the Association of American Geographers, 48*, 83-91.
Bowden, M.J. (1980) 'The Cognitive Renaissance in American Geography: The Intellectual History of a Movement', *Organon, 14*, 199-204.
Brookfield, H.C. (1962) 'Local Study and Comparative Method: An Example from Central New Guinea', *Annals of the Association of American Geographers, 52*, 242-54.

—— (1964) 'Questions on the Human Frontiers of Geography', *Economic Geography*, *40*, 283-303.

Brown, L.A. (1981) *Innovation Diffusion: A New Perspective*, Methuen, New York.

Brown, R.H. (1948) *Historical Geography of the United States*, Harcourt Brace, New York.

Brush, J.E. (1953) 'The Hierarchy of Central Places in Southwestern Wisconsin', *Geographical Review*, *43*, 380-402.

Bunge, W. (1979) 'Perspective on *Theoretical Geography*', *Annals of the Association of American Geographers*, *69*, 169-74.

Burton, I. (1963) 'The Quantitative Revolution and Theoretical Geography', *Canadian Geographer*, *7*, 151-62.

—— and Kates, R.W. (eds.) (1965) *Readings in Resources Management and Conservation*, University of Chicago Press, Chicago.

Buttimer, A. (1974) *Values in Geography*, Commission on College Geography, Resource Paper No. 24, Association of American Geographers, Washington DC.

——— and Seamon, D. (eds.) (1980) *The Human Experience of Space and Place*, St. Martins, New York.

Butzer, K.W. (1973) 'Pluralism in Geomorphology', *Proceedings of the Association of American Geographers*, *5*, 39-43.

—— (1980) 'Adaptation to Global Environmental Change', *Professional Geographer*, *32*, 269-78.

Calzonetti, F.J. and Eckert, M.S. (1981) *Finding a Place for Energy*, Resource Publications in Geography, Association of American Geographers, Washington DC.

Carson, R. (1962) *Silent Spring*, Houghton Mifflin, Boston.

Carter, D.B., Schmudde, T.H. and Sharpe, D. (1972) *The Interface as a Working Environment: A Purpose for Physical Geography*, Commission on College Geography, Technical Paper No. 7, Association of American Geographers, Washington DC.

Christaller, Walter (1966) *Central Places in Southern Germany*, Prentice-Hall, Englewood Cliffs, NJ.

Clark, A.H. (1954) 'Historical Geography' in P.E. James and C.F. Jones (eds.) *American Geography: Inventory and Prospect*, Syracuse University Press, Syracuse, 71-105.

—— (1968) *Acadia: The Geography of Early Nova Scotia to 1760*, University of Wisconsin Press, Madison.

Clarke, W.C. (1971) *Place and People: An Ecology of a New Guinea Community*, University of California Press, Berkeley and Los Angeles.

Colby, C.C. (1933) 'Centrifugal and Centripetal Forces in Urban Geography', *Annals of the Association of American Geographers, 23*,
1-20.

—— (1936) 'Changing Currents of Geographic Thought in America',
Annals of the Association of American Geographers, 26, 1-37.

Conzen, M.P. (1980) 'Historical Geography: North American Progress
in the 1970s', *Progress in Human Geography, 4*, 549-59.

Davis, W.M. (1909) *Geographical Essays*, Ginn and Company, Boston.

Detwyler, T.R. and Marcus, M.G. (1972) *Urbanization and Environment: The physical Geography of the City*, Duxbury, Belmont, Calif.

Dickinson, R.E. (1947) *City, Region, and Regionalism*, Routledge and
Kegan Paul, London.

Ehrenberg, R.E. (ed.) (1975) *Pattern and Process: Research in Historical Geography*, Howard University Press, Washington DC.

Entrikin, J.N. (1976) 'Contemporary Humanism in Geography', *Annals
of the Association of American Geographers, 66*, 615-32.

Fenneman, N.M. (1931) *Physiography of the Western United States*,
McGraw-Hill, New York.

—— (1938) *Physiography of the Eastern United States*, McGraw-Hill,
New York.

Finch, V.C. and Trewartha, G.T. (1936) *Elements of Geography*,
McGraw-Hill, New York.

Frazier, J.W. (ed.) (1982) *Applied Geography: Selected Perspectives*,
Prentice-Hall, Englewood Cliffs, NJ.

Furuseth, O.J. and Pierce, J.T. (1982) *Agricultural Land in an Urban
Society*, Resource Publications in Geography, Association of American Geographers, Washington DC.

Garrison, W.L. (1959-60) 'Spatial Structure of the Economy', *Annals
of the Association of American Geographers, 49*, 232-9, 471-82;
50, 357-73.

—— and Marble, D.F. (1957) 'The Spatial Structure of Agricultural
Activities', *Annals of the Association of American Geographers, 47*,
137-44.

Gersmehl, P.J. (1976) 'An Alternative Biogeography', *Annals of the
Association of American Geographers, 66*, 223-41.

Glacken, C.J. (1967) *Traces on the Rhodian Shore: Nature and Culture
in Western Thought from Ancient Times to the End of the Eighteenth Century*, University of California Press, Berkeley and Los
Angeles.

Golledge, R.G. and Rushton, G. (eds.) (1976) *Spatial Choice and
Spatial Behavior: Geographical Essays on the Analysis of Preferences*

and Perceptions, Ohio State University Press, Columbus.

Gould, P.R. (1979) 'Geography, 1957-1977: The Augean Period', *Annals of the Association of American Geographers, 69*, 139-50.

Guelke, L. (1974) 'An Idealist Alternative in Human Geography', *Annals of the Association of American Geographers, 64*, 193-202.

Hammond, K.A., Macinko, G. and Fairchild, W.B. (eds.) (1978) *Sourcebook on the Environment: A Guide to the Literature*, University of Chicago Press, Chicago.

Harris, C.D. (1943) 'A Functional Classification of Cities in the United States', *Geographical Review, 33*, 86-99.

—— (1954) 'The Geography of Manufacturing' in P.E. James and C.F. Jones (eds.), *American Geography: Inventory and Prospect, Syracuse University Press, Syracuse, 292-309.*

—— and Ullman, E.L. (1945) 'The Nature of Cities', *Annals of the American Academy of Political and Social Science, 242*, 7-17.

Harris, R.C. (1968) *The Seigneurial System in Early Canada*, University of Wisconsin Press, Madison.

Hart, J.F. (ed.) (1972) *Regions of the United States*, Harper and Row, New York.

—— (1975) *The Look of the Land*, Prentice-Hall, Englewood Cliffs, NJ.

—— (1982) 'The Highest Form of the Geographer's Art', *Annals of the Association of American Geographers, 72*, 1-29.

Hartshorn, T.A. (1980) *Interpreting the City: An Urban Geography*, Wiley and Sons, New York.

Hartshorne, R. (1939) 'The Nature of Geography: A Critical Survey of Current Thought in the Light of the Past', *Annals of the Association of American Geographers, 29*, 173-658.

Harvey, D. (1967) 'Models of the Evolution of Spatial Patterns in Human Geography' in R.J. Chorley and P. Haggett (eds.), *Models in Geography*, Methuen, London, 549-608.

—— (1973) *Social Justice and the City*, Johns Hopkins University Press, Baltimore.

Helburn, N. (1982) 'Geography and the Quality of Life', *Annals of the Association of American Geographers, 72*, 445-56.

Holcomb, H.B. and Beauregard, R.A. (1981) *Revitalizing Cities*, Resource Publications in Geography, Association of American Geographers, Washington DC.

Holz, R.K. (1973) *The Surveillant Science: Remote Sensing of the Environment*, Houghton Mifflin, Boston.

Hunter, J.M. (ed.), 1974, *The Geography of Health and Disease*, University of North Carolina Press, Chapel Hill.

Isaac, E. (1970) *Geography of Domestication*, Prentice-Hall, Englewood Cliffs, NJ.

Isard, W. (1956) *Location and Space Economy*, Wiley, New York.

James, P.E. (1935) *An Outline of Geography*, Ginn and Company, Boston.

—— and Jones, C.F. (eds.)(1954) *American Geography: Inventory and Prospect*, Syracuse University Press, Syracuse.

—— and Martin, G.J. (1978) *The Association of American Geographers: The First Seventy-Five Years, 1904-1979*, Association of American Geographers, Washington DC.

—— and —— (1979) 'On A.A.G. History', *Professional Geographer, 31*, 353-57.

Jordan, T.G. (1981) *Trails to Texas: Southern Roots of Western Cattle Ranching*, University of Nebraska Press, Lincoln.

King, L.J. (1976) 'Alternatives to a Positive Economic Geography', *Annals of the Association of American Geographers, 66*, 293-308.

—— and Golledge, R.G. (1978) *Cities, Space, and Behavior: The Elements of Urban Geography*, Prentice-Hall, Englewood Cliffs, NJ.

Knight, D.B. (1977) *A Capital for Canada: Conflict and Compromise in the Nineteenth Century*, Department of Geography Research Paper No. 182, University of Chicago, Chicago.

Kohn, C.F. (1954) 'Settlement Geography' in P.E. James and C.F. Jones (eds.), *American Geography: Inventory and Prospect*, Syracuse University Press, Syracuse, 124-41.

Kolars, J.F. and Nystuen, J.D. (1974) *Geography: The Study of Location, Culture, and Environment*, McGraw-Hill, New York.

Krumme, G. (1970) 'Location Theory' in P. Bacon (ed.), *Focus on Geography: Key Concepts and Teaching Strategies*, National Council for the Social Studies, Washington DC, 3-38.

Kuhn, T.S. (1962) *The Structure of Scientific Revolutions*, University of Chicago Press, Chicago.

Leighly, J. (1954) 'Innovation and Area', *Geographical Review, 44*, 439-41.

—— (1955) 'What has Happened to Physical Geography?', *Annals of the Association of American Geographers, 45*, 309-15.

—— (1979) 'Drifting into Geography in the Twenties', *Annals of the Association of American Geographers, 69*, 4-9.

Lentnek, B., Carmin, R.L. and Martinson, T.L. (eds.) (1971) *Geographic Research on Latin America: Benchmark 1970*, Ball State University Press, Muncie.

Ley, D. and Samuels, M. (eds.) (1978) *Humanistic Geography: Problems and Prospects*, Maaroufa, Chicago.

Lintz, J. and Simonett, D.S. (eds.) (1976) *Remote Sensing of Environment*, Addison-Wesley, Reading.

Lowenthal, D. (1961) 'Geography, Experience, and Imagination: Towards a Geographical Epistemology', *Annals of the Association of American Geographers, 51*, 241-60.

—— (ed.) (1967) *Environmental Perception and Behavior*, Department of Geography Research Paper No. 109, University of Chicago, Chicago.

—— *et al.* (1973) 'Report of the A.A.G. Task Force on Environmental Quality', *Professional Geographer, 25*, 39-47.

Manners, I.R. and Mikesell, M.W. (eds.) (1974) *Perspectives on Environment*, Commission on College Geography Publication No. 13, Association of American Geographers, Washington DC.

Marcus, M.G. (1979) 'Coming Full Circle: Physical Geography in the Twentieth Century', *Annals of the Association of American Geographers, 69*, 521-32.

Marsh, G.P. (1864) *Man and Nature; or, Physical Geography as Modified by Human Action*, Scribner, New York.

May, J.M. (1954) 'Medical Geography' in P.E. James and C.F. Jones (eds.), *American Geography: Inventory and Prospect*, Syracuse University Press, Syracuse, 452-68.

Mayer, H.M. (1954) 'Urban Geography' in P.E. James and C.F. Jones (eds.) *American Geography: Inventory and Prospect*, Syracuse University Press, Syracuse, 142-66.

—— and Kohn, C.F. (eds.) (1959) *Readings in Urban Geography*, University of Chicago Press, Chicago.

McCarty, H.H. and Lindberg, J.B. (1966) *A Preface to Economic Geography*, Prentice-Hall, Englewood Cliffs, NJ.

Meade, M.S. (ed.) (1980) *Conceptual and Methodological Issues in Medical Geography*, University of North Carolina Press, Chapel Hill.

Meinig, D.W. (1968) *The Great Columbia Basin: A Historical Geography, 1805-1910*, University of Washington Press, Seattle.

—— (1971) *Southwest: Three Peoples in Geographical Change, 1600-1970*, Oxford University Press, New York.

Merrens, H.R. (1964) *Colonial North Carolina: A Study in Historical Geography*, University of North Carolina Press, Chapel Hill.

Mikesell, M.W. (1978) 'Tradition and Innovation in Cultural Geography', *Annals of the Association of American Geographers, 68*,

1-16.

—— (1979) 'The A.A.G. at 75: Current Status', *Professional Geographer, 31*, 358-60.

—— (1981) 'Continuity and Change' in B.W. Blouet (ed.), *The Origins of Academic Geography in the United States*, Archon Books, Hamden, Conn. 1-15.

Mitchell, J.K. (1974) 'Natural Hazards Research' in I.R. Manners and M.W. Mikesell (eds.), *Perspectives on Environment*, Association of American Geographers, Washington DC, 311-41.

Morrill, R.L. (1981) *Political Redistricting and Geographical Theory*, Resource Publications in Geography, Association of American Geographers, Washington DC.

Myers, S.K. (1976) 'Geography in Action', *Geographical Review, 66*, 467-8.

Palm, R. (1981) *The Geography of American Cities*, Oxford University Press, New York.

Peet, R. (ed.) (1977) *Radical Geography: Alternative Viewpoints on Contemporary Social Issues*, Maaroufa, Chicago.

Pred, A. (1967-9) *Behavior and Location: Foundations for a Geographic and Dynamic Location Theory*, 2 vols., C.W.K. Gleerup, Lund.

Pyle, G.F. (1979) *Applied Medical Geography*, Wiley, New York.

Rappaport, R.A. (1969) *House Form and Culture*, Prentice-Hall, Englewood Cliffs, NJ.

Rooney, J.F., Zelinsky, W. and Lauder, D.R. (eds.) (1982) *This Remarkable Continent: An Atlas of United States and Canadian Society and Culture*, Texas A and M University Press, College Station.

Rose, H.M. (1971) *The Black Ghetto: A Spatial Perspective*, McGraw-Hill, New York.

—— (1976) *Black Suburbanization: Access to Improved Quality of Life or Maintenance of the Status Quo?* Ballinger, Cambridge, Mass.

Saarinen, T.F. (1969) *Perception of Environment*, Commission on College Geography Resource Paper No. 5, Association of American Geographers, Washington DC.

Sack, R.D. (1974) 'The Spatial Separatist Theme in Geography', *Economic Geography, 50*, 1-19.

Samuels, M.S. (1979) 'The Biography of Landscape: Cause and Culpability' in D.W. Meinig (ed.), *The Interpretation of Ordinary Landscapes*, Oxford University Press, New York, 51-88.

Sauer, C.O. (1925) 'The Morphology of Landscape', *University of Calif-*

ornia *Publications in Geography, 2*, 19-53.

—— (1941) 'Foreword to Historical Geography', *Annals of the Association of American Geographers, 31*, 1-24.

—— (1952) *Agricultural Origins and Dispersals*, American Geographical Society, New York.

—— (1971) *Sixteenth Century North America: The Land and People as Seen by Europeans*, University of California Press, Berkeley and Los Angeles.

Schmid, J.A. (1975) *Urban Vegetation: A Review and Chicago Case Study*, Department of Geography Research Paper No. 161, University of Chicago, Chicago.

Schwartzberg, J.E. (1978) *Historical Atlas of South Asia*, University of Chicago Press, Chicago.

Sopher, D.E. (1967) *Geography of Religions*, Prentice-Hall, Englewood Cliffs, NJ.

—— (ed.) (1980) *An Exploration of India: Geographical Perspectives on Society and Culture*, Cornell University Press, Ithaca.

Spencer, J.E. (1966) *Shifting Cultivation in Southeastern Asia*, University of California Publications in Geography 19, University of California Press, Berkeley and Los Angeles.

—— and Thomas, W.L. (1969) *Cultural Geography: An Evolutionary Introduction to Our Humanized Earth*, Wiley and Sons, New York.

Szymanski, R. and Agnew, J.A. (1981) *Order and Skepticism: Human Geography and the Dialectic of Science*, Resource Publications in Geography, Association of American Geographers, Washington DC.

Taaffe, E.J. (ed.) (1970) *Geography*, Prentice-Hall, Englewood Cliffs, NJ.

—— (1974) 'The Spatial View in Context', *Annals of the Association of American Geographers, 64*, 1-16.

Tatham, G. (1951) 'Environmentalism and Possibilism' in G. Taylor (ed.), *Geography in the Twentieth Century*, Philosophical Library, New York, 128-62.

Taylor, G. (1949) *Urban Geography*, Methuen, London.

Terjung, W.H. (1976) 'Climatology for Geographers', *Annals of the Association of American Geographers, 66*, 199-222.

Thomas, W.L. (ed.) (1956) *Man's Role in Changing the Face of the Earth*, University of Chicago Press, Chicago.

Trotier, L. (ed.) (1972) *Studies in Canadian Geography*, 6 vols, University of Toronto Press, Toronto.

Tuan, Yi-Fu (1974) *Topophilia: A Study of Environmental Perception, Attitudes, and Values*, Prentice-Hall, Englewood Cliffs, NJ.

Ullman, E.L. (1941) 'A Theory of Location for Cities', *American Journal of Sociology, 46*, 853-64.

Vale, T.A. (1982) *Plants and People: Vegetation Change in North America*, Resource Publications in Geography, Association of American Geographers, Washington DC.

Vance, J.E. (1977) *This Scene of Man: The Role and Structure of the City in the Geography of Western Civilization*, Harper and Row, New York.

Waddell, E. (1972) *The Mound Builders: Agricultural Practices, Environment, and Society in the Central Highlands of New Guinea*, University of Washington Press, Seattle.

Wagner, P.L. (1972) *Environments and Peoples*, Prentice-Hall, Englewood Cliffs, NJ.

—— and Mikesell, M.W. (eds.) (1962) *Readings in Cultural Geography*, University of Chicago Press, Chicago.

Ward, D. (1971) *Cities and Immigrants: A Geography of Change in Nineteenth-Century America*, Oxford University Press, New York.

Warkentin, J. (ed.) (1968) *Canada: A Geographical Interpretation*, Methuen, Toronto.

West, R.C. and Augelli, J.P. (1966) *Middle America: Its Lands and Peoples*, Prentice-Hall, Englewood Cliffs, NJ.

Wheatley, P. (1971) *The Pivot of the Four Quarters: A Preliminary Enquiry into the Origin and Character of the Ancient Chinese City*, Aldine, Chicago.

White, C.L. and Renner, G.T. (1945) *Human Geography: An Ecological Study of Society*, Appleton, Century, Crofts, New York.

White, G.F. *et al.*, (1958) *Changes in Urban Occupance of Flood Plains in the United States*, University of Chicago, Department of Geography Research Paper No. 57, Chicago.

—— and Haas, J.E. (1975) *Assessment of Research on Natural Hazards*, M.I.T. Press, Cambridge, Mass.

Whittlesey, D. (1954) 'The Regional Concept and Regional Method' in P.E. James and C.F. Jones (eds.), *American Geography: Inventory and Prospect*, Syracuse University Press, Syracuse, 19-69.

Wolpert, J. (1976) 'Opening Closed Spaces', *Annals of the Association of American Geographers, 66*, 1-13.

Wood, J.D. (ed.) (1981) *Rethinking Geographical Inquiry*, York University, Atkinson College, Department of Geography Monograph No. 11, Downsview, Ontario.

Zelinsky, W. (1973) *The Cultural Geography of the United States*, Prentice-Hall, Englewood Cliffs, NJ.

—— (1975) 'The Demigod's Dilemma', *Annals of the Association of American Geographers, 65*, 123-43.

—— (ed.) (1978) 'Human Geography: Coming of Age', *American Behavioural Scientist, 22*, 1.

10 HUMAN GEOGRAPHY IN THE NETHERLANDS
Chr. van Paassen

The Dutch geographical tradition differs considerably from that of its European neighbours in various respects. Most notable of the differences is that, since 1907, physical and human geography have followed entirely separate paths. Physical geography became almost identical to geomorphology. Human geography was identified explicitly as a social science, and has been called social geography since 1921.

Choice of the name 'social geography' represented a compromise between the two schools of human geography at the time, in Amsterdam and Utrecht. In Amsterdam, S.R. Steinmetz advocated 'sociography' as the 'grass roots' study of peoples and nations in their diversity, building on the common interests of differential sociology and human geography. This Amsterdam school of sociography functioned as the cradle for the development of social science in the Netherlands. In Utrecht, J.F. Niermeyer and his successor L. van Vuuren also regarded human geography as a social science, but they were adherents of the French school and much influenced by the work of Vidal de la Blache. They advocated a form of existential geography and studied the socio-economic way of life in the context of the so-called man-land paradigm.

Van Vuuren and H.N. ter Veen, who succeeded Steinmetz at Amsterdam in 1933, added a new dimension to social geography in the Netherlands. Both leaders adopted a policy-orientated approach, ter Veen mainly in regard to the colonisation of the IJsselmeerpolders and van Vuuren with especial reference to urban and regional planning. Thus research in human geography was strongly concentrated on problems of Dutch society. Many human geographers took up careers in government institutions such as the National Census Bureau and the national and provincial physical planning services (established in 1943). And in 1940 ter Veen became president of the newly-formed National Institute for Social Scientific Research (ISONEVO), which played a leading role in the development of geographical research.

1945-50

The immediate post-war years were marked by a demographic explosion, a great housing shortage, economic scarcity and an urgent need for economic recovery and industrialisation. Two ISONEVO conferences on societal planning and on the effects of population increases were set in this context, as was a programme of research into the 'birthrate problem'. The most important publication was a 1949 research report (much influenced by the British Barlow Report on the Distribution of Population, published in 1940) prepared at ISONEVO (and with a leading contribution from geographers) for the National Service for Physical Planning on the distribution of population.

Research done during the war resulted in other valuable publications during this period. Winsemius's (1945/9) research on the distribution of industry, undertaken for the Department of Economic Affairs, remained a reference work for many years. Keuning (1947) published the first modern regional geography of the Netherlands, and also produced the first central-functional classification of Dutch cities (Keuning, 1948). Christaller's central-place theory was applied to settlement policy in the Noordoostpolder, the new IJsselmeerpolder (Takes, 1948).

In the academic sphere, there was an interregnum at both Amsterdam and Utrecht. At the latter, J.O.M. Broek succeeded van Vuuren, and he in turn was replaced by A.C. de Vooys (1949-73). A year later, H.D. de Vries Reilingh (1950-71) was appointed at Amsterdam, and a new period of academic leadership was established. De Vries Reilingh, a pupil of Steinmetz, had written a PhD (1945) on the Folk High School movement, whereas that of de Vooys (1932), who studied at Utrecht, was on rural migration, innovative in its character by its fieldwork that followed the trend of American rural sociology. These two professors determined the course of research in human geography for the next decades. They were joined by H.J. Keuning, of the Utrecht school, who was appointed to the new university department at Groningen (1948-75).

During this period some of the best sociographers started their careers as sociologists, as a consequence of the establishment of that discipline in the universities. Six chairs were established and five were occupied by sociographers — several of them gifted geographers who therefore contributed to a brain drain from their original discipline, in which they were still active. Den Hollander (1946), for example, introduced a form of perception geography in his inaugural lecture entitled

'As others see us', and in 1947 he published a remarkable study of the European frontier region in Hungary.

A further chair of human geography was established at Wageningen, where van Vuuren had lectured before the war. It was occupied by E.W. Hofstee (1946-79), who established the internationally-known Wageningen school of rural sociology. His differential sociology was the continuation of the Steinmetz heritage *par excellence*, and retained its distinctive geographical features. His studies of socio-demographic processes and of modernisation processes in rural society belong to the highest level of Dutch social science, and his inaugural lecture on the causes of agrarian geographical differentiation was a topic of discussion for a long time (Hofstee, 1946).

The 1950s

This was a prolific decade, in which the work undertaken set trends for many years to come, both in academic circles and in policy-orientated research.

At Utrecht, a distinctive 'de Vooys school' evolved. His inaugural lecture (1950) introduced a variant of the French school, which continued to be influential: he proposed a socio-economic geography in which human existence and socio-economic structure constituted the focus, but the relationship with the physical environment was no longer the guiding principle. One characteristic of this work was his preference for small-scale research based on direct field observation, made manifest in studies of small cities (van der Kaa, 1959). His historical-geographical and socio-demographical interests were also remarkable and he became one of the experts in this area — along with Hofstee, who followed de Vooy's (1951) classic article on the mortality rate during the nineteenth century with his own large socio-demographic study of procreation (Hofstee, 1954), and sociologists like van Heek (1954). De Vooys (1959) also undertook similar small-scale field-work in underdeveloped regions of Spain and Greece. This led to a 'geography of development' and two of the students participating in that work now occupy chairs in the field — J. Hinderink at Utrecht since 1969 and J.M.G. Kleinpenning since 1972 at Nijmegen. Establishment of a new chair at the Catholic University of Nijmegan (first occupied by C. Cools: 1958-73) allowed the Utrecht school to expand its 'sphere of influence'.

The situation at Amsterdam was more difficult. De Vries Reilingh tried to occupy a location between empirical sociology and human geography. He had to defend sociography, which no longer had the empirical monopoly within the social sciences, against aggressive, young, sociologically-minded workers (one of his students – J.A.A. van Doorn – spoke of a sociographic 'intermezzo') as well as against Utrecht's orthodox-geographical repudiation of the Amsterdam heretics. He tried to defend the study of territoriality and rural culture against sociologists, seeking support in his concepts of territorial bond and community (de Vries Reilingh, 1964). His assistant, W.F. Heinemeijer, introduced human ecology, and advocated an urban sociography along those lines at a conference on Woudschoten in 1954, without much success. He put his ideas into practice in neighbourhood research and in a study of youth and leisure (Heinemeijer, 1964). Jolles's (1957) PhD on the declining birth-rate in Vienna was an original Amsterdam contribution to socio-demography.

The theoretical stance to sociography remained weak, however. A group of young sociologists, led by J.A.A. van Doorn, established a new sociological journal – *Sociologische Gids* – in 1953. He and Lammers (1958) launched their detailed and principal attack on sociography in this journal: the defence was honest, but weak. Meanwhile, Heinemeijer went his own way with field-work in Morocco, the first steps in the development of a modern political geography and also of a geography of development, Amsterdam-style.

At Groningen, Keuning continued his prolific output of articles on the geography of the Netherlands. His masterpiece was *Mosaic of Functions* (1955), a cross-sectional, historical geography of the country.

During this decade there was a stimulating climate for policy-orientated geography, with many studies of local and regional socio-economic structure. ISONEVO, together with other national institutes, organised a large research project into problem areas of the Netherlands (1953-6). It also organised three major conferences: one on social aspects of the Zuiderzee reclamation (1952); one on policy-orientated sociography, at which W. Steigenga discussed urban research (1952); and one about social integration in new neighbourhoods (1955), at which van Doorn (1955) attacked the ideology of the neighbourhood concept.

Policy-orientated geographical research flourished in many government and semi-governmental institutions. There was an important research centre associated with the Rijksdienst voor het Nationale Plan, for example. There were similar research centres at the provincial level, including a large and productive one in North-Holland led by G.J. van den Berg, who introduced new methods of public participation in the planning process. He also developed new concepts regarding the process of urbanisation, presented at a conference on New Towns at the Netherlands Institute of Town Planning and Housing (1957) and at a conference (also in 1957), convened by a federation of Dutch social science students, on the Metropolis. The report on New Towns soon became a classic, and together with Steigenga's ideas greatly influenced national physical planning policy. (The first report – *Nota inzake de Ruimtelijke Ordening in Nederland* – appeared in 1960.) Further, in 1955 Steigenga published his classification of towns based on socio-economic vulnerability, and in 1958 the National Census Bureau published a classification (by a Utrecht geographer, J. Schmitz) of municipalities, based on their degree of urbanisation.

Policy-orientated geography also flourished in the universities. A detailed study of the Haarlemmermeer was published at Utrecht in 1955, introducing a spatial approach, strongly idiographic and situational in its character, to investigations of the impact of suburbanisation (Kouwe and Wissink, 1955). At the same time, interdisciplinary communication in the field of policy-orientated research was strengthened. Essential to this was the series of Woudschoten conferences, organised by ISONEVO and a new association of applied social researchers; these provided a forum for social scientists and provided channels of communication for geographers, despite the antagonism with sociologists.

Other fields of geography were relatively unpopulated. In the theoretical field there was silence except for two pieces – van Paassen's (1957) PhD on the classical tradition and de Jong's (1962) analysis of the chorological approach of Hettner and Hartshorne. Historical geography flourished through the works of Keuning and Elizabeth Gottschalk.

The 1960s

This was a decade that experienced an extraordinary increase in staff and student numbers. It ended in the well-known period of student unrest, symbolised by student occupation of the administrative centre of the University of Amsterdam (the 'Maidenhouse') in the centre of the city.

For academic geography , expansion in the 1960s was expansion of the Utrecht school and its influence. A new chair was established there – the first in the Netherlands – for the human geography of underdeveloped regions (J.J. Hanrath, 1965-9; J. Hinderink, 1969-), and geography was established at the Protestant University in Amsterdam (The Free University); the occupant of the chair, M. Heslinga (1967-), soon organised a large department along Utrecht lines. Elsewhere, a second chair of human geography was established at Groningen and occupied by R. Tamsma (1966-), whose position at Rotterdam was filled by an Englishman, P.R. Odell (1968-80); a second chair of human geography was created at Amsterdam, and filled by Heinemeijer (1968-); and, also at Amsterdam, the only chair of historical geography was installed, to be occupied by Elizabeth Gottschalk (1964-77).

Some of the best geographers became professors of physical planning (called 'planologie', because of the absence of chairs in policy-orientated geography). W. Steigenga was appointed at Amsterdam (1962-74), for example, and became the country's leading *planeloog*. A young urban geographer, G.A. Wissink, from the National Planning Service, who had just finished his PhD on urban fringe areas in the United States (Wissink, 1962) was appointed to Nijmegen (1962-), and in 1968 the experienced policy-orientated geographer G.J.A. van den Berg (who had taught with van Paassen at Utrecht) was appointed at Groningen; van Paassen left the geographical institute at Utrecht for a chair of planologie at the same university (1967-73).

Planologie and human geography continued to diverge in their content. Steigenga continued to write stimulating articles on urbanisation and suburbanisation and also produced a book on modern planologie (1964) with a definite geographical character, but in his inaugural lecture (Steigenga, 1963) advocated a new course on 'Socio-Spatial Construction'. This approach was extended later by van den Berg (1981).

At Amsterdam, a new chair of sociography was established in 1961 in the faculty of social and political sciences, and the subject was implemented as applied sociology. De Vries Reilingh's role was now limited to the faculty in which geography was taught, and as a consequence the need for a 'geographisation' of sociography was strengthened.

In substance, the relative characters of geography and differential sociology remained as in the previous decade. Several large PhD studies in socio-demography were reported (e.g. ter Heide, 1965, on internal migration, and Buissink, 1970, on Dutch regional variations in fertility rates), and emigration was a subject of considerable interest. Hofstee (1962, 1963) and de Vooys (1964) continued their work in this field, and there was a fierce polemic involving Hofstee and van Heek (1963) over the importance of the religious factor (compare Knippenberg, 1980, and Verduin, 1972). Several rural sociological studies were published by the Hofstee school (e.g. Bergsma, 1963; Lijfering, 1968). One was on the planning of the IJsselmeerpolders and was written by a geographer from Utrecht, Constandse (1960), who undertook part of his graduate work at Wageningen; it contained an analysis of the relationship between rural culture and settlement pattern, and implemented the concept of livability.

Studies of rural geography outside the Netherlands were of an entirely different character. Van Hulten (1964) investigated agricultural collectivisation in Poland, for example, and Tamsma's (1966) study of the Moshav Ovdiem colonies in Israel was remarkable for its consistent functional approach to the physical environment.

In spite of the large volume of work on rural geography (still a favourite theme), innovation in Dutch geography at this time was mainly in urban studies, and closely connected to policy-orientated work. Indeed, 1960-70 might be regarded as the decade of urban geography. In 1966, for example, the government published its Second Report on Physical Planning and the Royal Dutch Geographical Society (KNAG) held a conference on it. In the same year, a conference was held in Amsterdam on Urban Core and Inner City, attracting many well-known geographers from abroad, and the Woudschouten Conference on problems of social organisation included a trend-setting lecture (Kouwe, 1970) on the organisational structure of the city. The human geography section of KNAG organised two other conferences, one on the regional concept in general and one on the city-region in particular (Lambooy, 1970).

Interest in urban geography developed from 1962 on. In that year, students at Utrecht chose 'the city in a new setting' as the topic for their lustrum conference, and new viewpoints were presented there by Hoekveld, Kouwe and Wissink. An article by van Paassen (1962) on the ecological complex aroused interest through its approach directed at the processes of geographical structuring leading to both integration and differentiation in an urban complex. The article was connected to policy-orientated geography, for it was a geographer's response to an inter-disciplinary, applied research project sponsored by ISONEVO (now called the Foundation for Inter-University Social Research — SISWO) on the problems of the inner city. That project involved several university geography institutes, including Amsterdam, where Heinemeijer continued his socio-ecological research in the inner city.

Heinemeijer (1963) published the first review of urban ecology in the Netherlands, at last bringing about a meeting of sociography and social ecology. His main contribution was his work on the attraction of the urban core (Heinemeijer, 1968), which contains an interesting typology of urban residents with regard to their 'urban attitude' (a topic taken up in the PhD of his assistant, van Engelsdorp Gastelears, 1980). Urban geography was also being developed at the Free University, where Hoekveld, who had made a valuable study of a Dutch suburb (1964), introduced a systems approach based on some elements of van Paassen's ecological complex concept; Lambooy's (1969) inaugural lecture at the Free University took up the same theme.

Heinemeijer's school also enriched the subdiscipline of political geography during this period. Influenced by, among others, the American political scientist Karl Deutsch, Heinemeijer undertook his PhD study on mobilisation and integration in the new state of Morocco; this was the first such study in the Netherlands (Heinemeijer, 1968). With his assistant, van der Wusten, he also developed political geography along lines suggested by the work of the Norwegian, Stein Rokkan. Another major contribution, linking political and cultural geography, was the PhD by Heslinga on 'The Irish Border as a Cultural Divide' (1962).

Institutionally, Dutch geography was reorganised in 1967 with the merger of three separate societies into the one, the reformed Royal Dutch Geographical Society. The country's two main journals – *Tijdschrift voor Economische en Sociale Geografie (TESG)* and *Tijdschrift van het Koninklijk Nederlands Aardrijkskundrig Genootschap (TAG)* – were also merged, taking the title of the former. The *Geografisch Tijdschrift* was reissued in a new form, and acquired a more general character. Finally, during the 1960s the Atlas of the Netherlands project (1963-77) was launched.

From 1970

During this most recent period human geography achieved its greatest expansion and highest level of productivity. It has also been a period of diversification and of 'changing of the professorial guard' in most universities.

At Amsterdam, personnel changes at the top saw Heinemeijer replace de Vries Reilingh; he was assisted by van Paassen (1973-82), who returned to geography from planologie. At Groningen and Nijmegen respectively, Keuning was replaced by Kouwe (1974) and Cools by Kleinpenning (1973). A further professorial appointment at Groningen was of P. Lukkes, whose inaugural lecture (1974) introduced the concept of 'regiology'. (Keuning's inaugural there in 1949 was on regionalism.)

At Utrecht, leadership passed for the first time to the younger generation of geographers (born in the 1940s); M. de Smidt (1978-) is a specialist on the geography of labour markets (de Smidt, 1975), and J. van Ginkel (1979-) produced a two-volume PhD thesis (1979) on suburbanisation and residential environments. The latter piece began as part of an applied geography project on the central 'green heart' of Randstad Holland (Ottens, 1976), but was nonetheless remarkable for its theoretical-methodological content. Borchert, Hauer, Ottens and de Smidt all participated in that project, and the contribution of H.E.L. Ottens marked the introduction of the systems approach at Utrecht.

There was a further substantial increase in the number of chairs at the end of the decade. Four chairs in geography were established at the Free University (including one for the geography of underdeveloped areas, occupied by an urban geographer, G.A. de Bruyne); two were created at Nijmegen (one in economic geography, occupied by E. Wever whose PhD was on the petrochemical industry, and one in urban geography, occupied by J. Buursink, a pupil of Keuning and an expert on central-place theory: Buursink, 1971); and one was established at Groningen, occupied by W.J. van den Bremen. Two chairs in planologie were established at the Free University, one of them occupied by J. Buit, a geographer at the Utrecht planologie institute. And a related field was opened up by the appointment of a student of the urban habitat (see Grunfeld, 1970) to a chair in the applied sociology of urban and regional planning at the University of Tilborg.

Economic geography began to flourish. Lambooy, who was appointed to the economics faculty at Amsterdam, wrote a text in collaboration with A.G.M. Jansen and H.W. ter Hart, which became a standard in the field. Jansen published widely on industrial development and the locational behaviour of multinational firms in the Netherlands, and collaborated with de Smidt (1974, 1979) in a book on the former topic; ter Hart's (1979) PhD was on the location of top management. Lambooy combined geography with regional economics, with particular emphasis on institutional factors and externalities (1973); in Groningen, Lukkes developed applied regional economic research; and in Utrecht and Nijmegen de Smidt and Wever investigated mobility and the relocation of industrial plants and offices.

The 1970s saw many new textbooks, with both the Hoekveld and Utrecht schools particularly productive; Kleinpenning proved to be a prolific author of texts on the Third World. Hoekveld (1971a, 1971b) also paid much attention to the conceptual structure of geography.

Productivity and diversification were typical of this period, as clearly illustrated in the new series of national geographical study-conferences, held in Utrecht (1977), Amsterdam (1980) and Nijmegen (1982). It was also the period when the 'quantitative revolution' belatedly reached Dutch geography, and generalisation based on an analytical approach, involving systems and models, became paramount. (The 'spatial paradigm' had few roots in Dutch geographical tradition.) The first empirical applications came from the Amsterdam school and were much influenced by sociological manuals – J. Galtung's *Theory and Methods of Social Research* (1970) was especially important in this respect. In 1971, SISWO and KNAG convened a seminar on 'Mathematical Methods in Human Geography'; this was led by J. Hauer, who published the first text on quantitative research methods (Hauer and van der Knaap, 1973). Hauer was appointed (in 1975) to the first chair in the Netherlands established for research methodology in geography; a second appointment (F.M. Dieleman) was made at the Free University in 1978. The technical advances were soon apparent in PhD theses such as those by van der Knaap and Dieleman (both in 1978), Dietvorst (1979), Timmermans (1979), Floor and de Jong (1981).

The substantive content of much human geography in the 1970s followed lines established in earlier decades. Policy-orientated work continued to be conducted in close collaboration with government institutions, and included Steigenga's masterpiece (*Globale Visie*) on the Utrecht conurbation. The third report of the National Service of Physical Planning (1975) also stimulated much discussion, and was the subject of two symposia organised by KNAG.

Historical geography continued to flourish at Amsterdam, under the leadership of Elizabeth Gottschalk who published three lengthy, detailed studies of storm floods and fluvial inundations (1971, 1975, 1977); she was succeeded by her assistant, G. Borger, in 1980. Amsterdam was also a centre of research into ethnic minorities. Van Amersfoort (1974) undertook a pioneering PhD study of immigration and the development of minorities in the Netherlands, and Dostal and Knippenberg (1979) studied Russification in the Soviet Union.

In the study of underdeveloped regions, Hinderink, in co-operation with a Turkish sociologist, M. Kiray (1970), undertook field-work in Turkey. He then switched his interests, first to Africa, where he conducted a study of the Cape Coast (Hinderink and Sterkenburg, 1975), and then to Indonesia. De Bruyne, of the Free University, continued his urban geographical research in Paramaraibo, and Kleinpenning's main interest focused on Latin America. The Amsterdam school had its laboratory in Morocco (de Mas, 1981), where it operated in the context of a government-sponsored project (REMPLO) into topics such as the impact of migrant labour (Heinemeijer *et al.*, 1977). The importance of this focus was shown at the 1978 conference in Utrecht, when an entire section was devoted to 'Spatial inequality and government policy in developing countries' (TESG, 1978, Parts 1-2, 1-104).

The development of political geography at Amsterdam culminated with the first general text in the field published in the Netherlands (Amersfoort, Heinemeijer and Wusten, 1981). Van der Wusten (1977) wrote an original PhD study on Irish resistance to political integration and also provided a Dutch contribution to a study of the roots of European fascism (1980). In 1982 he published a study of 'Finlandisation' as a precursor of 'Hollanditis' (Kwaasteniet and van der Wusten, 1982).

Urban geography continued to predominate, however, with Amsterdam developing social ecology (Cortie, van Engelsdorp Gastelaars, Ostendorf) and the other institutes mainly concentrating on economic and labour market studies. Hoekveld's group at the Free University spanned both areas, and they published a valuable book on the Dutch city (Deurloo, Dieleman and Hookveld, 1981). In most of the departments there was a tendency to concentrate on their local environments. This was certainly the case with the two in Amsterdam. At the University of Amsterdam, for example, the long tradition of work on segregation is reflected by van Engelsdorp Gastelaar's (1980) PhD in social ecology and by the joint conference (KNAG and the Dutch Sociological Association) held there (Blauw and Pastor, 1980). Work was also developed on outdoor activities, influenced by Hagerstrand's time geography; time budget and human activity analysis was undertaken in co-operation with Velibor Vidakovic of the Amsterdam Board of Municipal Works. At the Free University, Hoekveld's team published a book on Amsterdam in 1980. The Utrecht school continued with its large-scale research projects (e.g. STEPRO, 1978), and in Rotterdam a large study culminated in a detailed report on social segregation there (Mik, 1980). During this decade, too, the National Census Bureau distributed research projects related to the 1971 census of population among the social science departments. One of the monographs produced was on a new typology of municipalities (van Engelsdorp Gastelaars, Ostendorf and de Vos, 1980); another was on regional labour markets (Schuurmans, 1981).

The 1970s also witnessed a period of strong ideological allergy and there were fierce science-political discussions associated with a revival of theoretical interest. Van Amersfoort (1972) initiated this with a paper on 'critical geography', and a new journal, *Zone*, with a strong Marxist bias was established by students, mainly from Nijmegen. Also at Nijmegen, involved in the spiritual and political turmoil within the Roman Catholic world, a conference on 'critical geography' was organised in 1973; the 'Positivismusstreit' was discussed, and Habermas and Popper compared. And in 1976, the KNAG organised a conference on theoretical reflections on the course of geography, where Jansen (1976) introduced the concept of policy-orientated geography and van Paassen (1976) outlined his existentialist-anthropological viewpoint. It was also Jansen (1977; see also Lambooy, 1978) who criticised work on centrality and hierarchies by Buursink (1971, 1977) and Buit and Nozeman (1976), and who advocated a phenomenological approach (Jansen, 1980); he applied his alternative, phenomenologically and policy-orientated, approach in a study of Amsterdam (Jansen, 1982).

Retrospect and Prospect

Until the 1970s the bipolar structure based on Amsterdam and Utrecht dominated Dutch geography. The 'new professional guard' did not change that and the weight of a few personalities remained substantial, despite the challenges to academic traditions from changes within society and the competition from other rapidly expanding social sciences. Amsterdam had to defend sociography as a legitimate alternative to sociology and Utrecht had to adapt French traditional geography to the study of an urban society. Both chose to cultivate factionalism and remain within their institutional traditions, however; Amsterdam remained in close proximity to sociology while Utrecht continued to prefer the study of economic phenomena. In this, personal interests and biases were important. At Amsterdam, the urge for 'geographicalisation' of sociography was strongly personally-based, as was the prevalent rural and cultural interest; a spatial viewpoint was adopted, but operated within the context of concepts derived mainly from sociology. At Utrecht, on the other hand, geographical orthodoxy was favoured, although historical and demographic interests and a preference for small-scale field-work, at home and abroad, gave rise to new substantive work.

It remains remarkable that the new developments in the Anglo-Saxon and Scandinavian geographical professions had no impact on

Dutch geographers at the time, despite the expansion of numbers and the establishment of new departments. Amsterdam's international social network consisted mainly of sociologists and cultural anthropologists, whereas Utrecht's international network was small and its appreciation of new geographical ideas slight.

New to this period in the Netherlands was the explosion of work in policy-orientated geography linked outside the walls of academia, especially with public urban and regional planning bodies. A new, large group of applied geographers came to the fore. They had to co-operate with colleagues from other disciplines, and were thus open to new ideas. As a result, they operated as the innovative force within geography, stimulating work in urban geography and in the study of relevant theory.

The relationship between academic and applied geography suffered tensions, however. Newly-created chairs of planologie were occupied by geographers, but which context would they prefer? Would geography continue to offer the main conceptual and instrumental inputs, or would a new branch of planning-orientated 'clinical' and constructive social science arise, with roots elsewhere?

The 1970s brought a kind of caesura to the history of Dutch human geography. A few personalities no longer dominated; a new generation took over control; and an even larger group of yet-younger staff implemented the international methodological revolution. The spatial paradigm took hold. There was a convergence between Amsterdam and Utrecht but a radical divergence of planologie and geography; planologie chose social construction, and geography spatial analysis. Specialisation within geography developed rapidly, with links being forged with other relevant social sciences, in great contrast to the preceding decade. New subjects and problems were taken up; a new explosion of policy-orientated work focused on development issues took place; and in many textbooks Dutch geographers codified a human geography in line with international trends.

These new developments were accompanied by critical reflections based on both Marxist-structuralist ideas and a phenomenologically-inspired anthropological approach. The spatial paradigm was questioned because experience indicated a growing gap between abstract formalism and spatial content and relevance.

Further challenges in the near future will have to be responded to under evidently unfavourable conditions. Academic research funds will be reduced radically, and public administrative centralisation will further standardisation and pragmatism in social research. The measures of mass will dominate, and the scope for diversity and critical reflection might be reduced. Issues of equality and inequality will undoubtedly flare up.

Human geography might well be torn into two directions. One will reflect the 'hard' course of analytical research directed to studies of the spatial dimensions of societal systems and of large-scale organisations; the other will be the 'soft' course, of qualitative, situational research focused on the many and pluriform realities of man. (There may be an intermediate course, for handling the problems of territorial co-existence and environmental diversification.) Geography should remain a battleground for different approaches. It should, however, pay more attention to institutional power and culture. Recent developments of an institutionally-directed economic geography and of political geography offer new prospects, as do a few experiments in qualitative research. Out of these, a new, more relevant policy-orientated geography, might recover some of the lost ground and the Dutch tradition of 'social geography' gain a new impetus.

References

The following abbreviations are used:
GT　　= *Geografisch Tijdschrift*
TAG　= *Tijdschrift van het Koninklijk Nederlands Aardrijkskundrig Genootschap*
TE(S)G = *Tijdschrift voor Economische (en Sociale) Geografie.*

Amersfoort, J.M.M. van (1972) 'Kritische Theorie en geografiebeoefering', *GT, 6*, 1-10.
—— (1974) *Immigratie en Minderheidsverming*, Samsen, Alphen aan den Rijn.
—— Heinemeijer, W.F. and Wusten, H.H. van der (eds.) (1981) *Een Wereld van Staten*, Samson, Alphen aan den Rijn.
Atlas van Nederland (1963-77) Staatsdrukkerij, The Hague.
Berg, G.J. van den (1957) 'Op Zoek naar een Kader' en 'De Mensen en de Nieuwe Steden' in *Nieuwe Steden*, Praeadviezen, Samson, Alphen aan den Rijn.
—— (1966) 'Over de Bruikbaarheid van het Regio-Begrip in de Geografische Planning', *TAG, 83*, 24-39.
——— (1981) *Inleiding tot de Planologie*, Samson, Alphen aan den Rijn.
Bergsma, R. (1963) *Op Weg naar een Nieuw Cultuurpatroon*, Van

Gorcum, Assen.

Blauw, P.W. and Pastor, C. (eds.) (1980) *Soort bij Soort,* Beschouwingen over Ruimtelijke Segregatie als Maatschappelijk Probleem, Van Loghum Slaterus, Deventer.

Bruyne, G.A. de (1976) *Bijdragen tot de Sociale Geografie van de Ontwikkelingslanden*, Romen, Bussum.

Buissink, J.D. (1970) *De Analyse van Regionale Vershillen in de Huwelijksvruchtbaarheid*, Delftse Uitgeversmaatschappij, Delft.

Buit, J. and Nozeman, E.F. (1976) *De Gewenste Spatiering van het Voorzieningenapparaat*, Geografisch en Planologisch Instituut, Vrije Universiteit, Amsterdam.

Buursink, J. (1971) *Centraliteit en Hierarchie*, Van Gorcum, Assen.

―― (1977) *De Hierarchie van Winkelcentra*, Geografisch Instituut, Universiteit, Groningen.

Constandse, A.K. (1960) *Het Dorp in de IJsselmeerpolders*, Tjeenk Willink, Zwolle.

Deurloo, M.C., Dieleman, F.M. and Hookveld, G.A. (1981) *Zicht op de Nederlandse Stad*, Romen, Haarlem.

Dieleman, F.M. (1978) *Een Analyse van Spreidingspatronen van Vestiging en van Werkgelegenheidsgebieden in Tilburg en Eindhoven*, Krips Repro, Meppol.

Dietvorts, A.G. (1979) *Telefoonverkeer en Economische Structuur in Nederland*, Nijmeegse Geografische Cahiers, no. 17, Nijmegen.

Doorn, J.A.A. van (1955) 'Wijk en Stad, Reele Integratiekaders?' in *Prae-adviezen voor het Congres over Sociale Samenhangen in Nieuwe Stadswijken*, ISONEVO, Amsterdam, 60-85.

―― and Lammers, C.J. (1958) 'Sociologie en Sociagrafie', *Sociologische Gids, 6*, 49-78, 141-2.

Dostal, P. and Knippenberg, H. (1979) 'The "Russification" of Ethnic Minorities in the USSR', *Soviet Geography, 20*, 197-219.

Engelsdorp Gastelaars, R.E. van, Ostendorf, W.J.M. and Vos, S. de (1980) *Typologieen van Nederlaandse Gemeenten naar Stedelijkheidse graad*, Monografie Volkstelling 1971, 15B, Staatsuitgeverij, The Hague.

―― (1980) *Niet Elke Stadsbewoner is een Stedeling*, Sociaal-Geografisch Instituut, Publicatie no. 6, Universiteit van Amsterdam.

Floor, J. and Jong, Th. de (1981) *Ontwikkeling en Toetsing van een Woonallocatie-Model*, 2 vols., Utrechtse Geografische Studies no. 22, Utrecht.

Ginkel, J.A. van (1979) *Suburbanisatie en Recente Woonmilieus*, Utrechtse Geografische Studies, no. 15 and 16, Utrecht.

Globale Visie (1970) Hoofdlijnen van de Inrichting van de Ruimte in Midden-Utrecht, 2 vols., Utrecht.

Gottschalk, M.K. (1971, 1975, 1977) *Stormvloeden en Rivieroverstremingen in Nederland*, 3 vols., Van Gorcum, Assen.

Greenman, Sj. (1956) *Uitdijende Werelden*, Inaugural Lecture, Leiden, Van Gorcum, Assen.

Grunfeld, F. (1970) *Habitat and Habitation*, Samsen, Alphen aan den Rijn.

Hart, H.W. ter (1979) *Vestigingsplaatsaspecten en Topmanagement*, Krips Repro, Meppel.

Hauer, J. and Knaap, G.A. van der (1973) *Sociale Geografie en Ruimtelijk Onderzeek*, Universitaire Pers, Rotterdam.

Heek, F. van (1954) *Het Geboorteniveau van de Nederlandse Rooms-Katholieken*, Stenfert Kroese, Leiden.

—— (1963) 'Het Nederlandse Geboortspatroon en de Godsdienstfactor gedurende de Laatste Halve Eeuw', *Mens en Maatschappij, 38*, 81-103, 257-68.

Heide, H. ter (1965) *Binnenlandse Migratie in Nederland*, Staatsdrukkerij, The Hague.

Heinemeijer, W.F. (1963) 'De Sociologie van de stad' in J.S. van Hessen (ed.), *Wegwijzer in de Sociologie*, De Bussy, Amsterdam.

—— (1964) *Buurt, Jeudg en Vrije Tijd*, Gemeentelijk Bureau voor de Jeugdzorg, Amsterdam.

—— Hulten, M. van and Vries Reilingh, H.D. de (1968) *Het Centrum van Amsterdam*, Polak en van Gennep, Amsterdam.

——- (1968) *Nationale Intergratie en Regionale Diversiteit*, De Bezige Bij. Amsterdam.

—— et al. (1977) *Partir pour Rester*, Incidences de l'émigration ouvrière à la Campagne Marocaine, NUFFSC/JMWOO, The Hague.

—— (1977) 'De Sociaal-Geografische Bemoeienis met Stad on Stedelijk Systeem', *GT, 11*, 259-72.

Heslinga, M.W. (1962) *The Irish Border as a Cultural Divide*, Van Gorcum, Assen.

—— (1978) 'Sociografie versus Sociale Geografie' in V. Bovenkerk *et al.* (eds.) *Toen en Thans*, Ambo, Baarn, 53-67.

Hinderink, J. and Kiray, M. (1970) *Social Stratification as an Obstacle to Development, a Study of Four Turkish Villages*, Praeger Publishers, New York.

—— and Sterkenburg, J. (1975) *Anatomy of an African Town*, A Socio-Economic Study of Cape Coast, Ghana, Geografisch Instituut, Utrecht.

Hoekveld, G.A. (1964) *Baarn*, Bosch en Keuning, Baarn.

—— (1971a) 'De Geografische Beschouwingswijze' in G.A. de Bruijne, G.A. Hoekveld, P.A. Schat, *Geografische Verkenningen*, Romen, Roermond, 11-35.

—— (1971b) *Geleding en Ontleding van de Stad*, Inaugural Lecture, Vrije Universiteit, Amsterdam, Kek, Kampon.

Hofstee, E.W. (1946) *Over de Oorzaken van de Verscheidenheid in de Nederlandse Landbouwgebieden*, Inaugural Lecture, Wageningen, Wolters, Groningen.

—— (1954) 'Regionale Verscheidenheid in de Ontwikkeling van het Aantal Geboerten, in Nederland in de Tweede Helft van de 19de Eeuw' in *Akademie-dagen VII*, Amsterdam, 59-106.

—— (1962) 'De Groei van de Nederlandse Bevolking' in A.N.J. den Hollander (ed.), *Drift en Koers*, Van Gorcum, Assen, 13-84.

—— (1963) 'Het proces der Geboertendaling in Nederland 1850-1960', *Mens en Maatschappij, 38*, 104-33, 269-77.

—— (1978) *De Demografische Ontwikkeling van Nederland in de Negentiende Eeuw*, Van Loghum Slaterus, Arnhem.

Hollander, A.N.J. den (1946) *Het Andere Volk*, Inaugural Lecture, Amsterdam, Sijthoff's Uitgeversnaatschappij, Leiden.

—— (1947) *Nederzettingsvormen en – problemen in de Grote Hongaarse Laagvlakte*, Meulenhoff, Amsterdam.

Hulten, M.H.M. van (1964) *De Collectivisatie van de Landbouw in de Volksrepublick Polen*, Drukkerij H.Y. Koersen en Zonen, Amsterdam.

Jansen, A.C.M. and Smidt, M. de (1974, 1979) *Industrie en Ruimte*, Van Gorcum, Assen.

—— (1976) 'On Theoretical Foundations of Policy-Oriented Geography', *TESG, 67*, 342-51.

—— (1977) *Over Armoede van Distributie-Planologisch Onderzoek*, E.G.I. Paper 14, Economisch Geografisch Instituut, Universiteit van Amsterdam.

—— (1980) 'Een Fenomenologische Orientatie in de Ruimtellijke Wetenschappen', *GT, 14*, 99-111, 324-9.

—— (1982) *Ondernemers in Camperstraat en Omgeving*, E.G.I. Paper 21, Economische Geografisch Instituut, Universiteit van Amsterdam.

Jolles, H.M. (1957) *Der Geburtenrueckgang in Wien*, Van Gorcum, Assen.

Jong, G. de (1962) *Chorological Differentiation as the Fundamental Principle of Geography*, Wolters, Groningen.

Kaa, D.J. van der (1959) 'Over het Verband tussen Bevolkingsgroei en

Centrumfunctie van Oostburg', *TAG, 76,* 363-78.

Keuning, H.J. (1947) *Het Nederlandse Volk in Zijn Woogebied* (revised edition 1965), Leopold's Uitgeversmaatschappij, The Hague.

—— (1948a) 'Proeve van een Economische Hierarchie van de Nederlandse Steden', *TEG, 39,* 566-81.

—— (1948b) *Regionalisme*, Inaugural Lecture, Groningen, Wolters, Groningen.

—— (1950) 'Eon Typologie van de Nederlandse Steden', *TESG, 41,* 187-206.

—— (1955) *Mozaiek der Functies*, Leopold's Uitgeversmaatschappij, The Hague.

Knaap, G.A. van der (1978) *A Spatial Analysis of the Evolution of an Urban System*, The Case of the Netherlands, Utrecht.

Knippenberg, H. (1980) 'De Demografische Ontwikkeling van Nederland sedert 1800', *GT, 14,* 54-74.

Kouwe, P.J.W. and Wissink, G.A. (1955) *De Haarlemmermeer*, Plattelandsproblemen in de Randstad Holland, Van Gorcum, Assen.

—— (1970) 'De Stad als Organisatiekader' in A. Bours and J.G. Lambooy (eds.), *Stad en Stadsgewest in de Ruintelijke Orde*, Van Gorcum, Assen, 163-74.

Kwaasteniet, M. de, Wusten, H. van der (1982) *Finlandisering, de Voerloper van Hollanditis*, Sociaal-Geografisch Instituut, Publicatie No. 7, Universiteit van Amsterdam.

Lambooy, J.G. (1966) 'Het Begrip Regio in de Geografische Theorie en Methode', *TAG, 83,* 15-23.

—— (1969) *Het Geografisch Systeem en de Groeipooltheorie*, Inaugural Lecture, Vrije Universiteit, Amsterdam, Van Gorcum, Assen.

—— (1970) 'Stad en Stadsgewest in het Perspectief van Hierarchie en Complementariteit' in A. Bours and J.G. Lambooy (eds.), *Stad en Stadsgewest in de Ruimtelijke Orde*, Van Gorcum, Assen, 270-91.

—— (1972, 1981) *Economie en Ruimte*, Van Gorcum, Assen.

—— (1973) 'Economic and Geonomic Space', *Papers of the Regional Science Association*, vol. 31, 145-58.

—— (ed.) (1978) *Beleidsgeorienteerd Onderzoek ter Diskussie* (see Jansen, 1977), E.G.I. Paper 17, Economische-Geografisch Instituut, Universiteit van Amsterdam.

Lijfering, J.H.W. (1968) *Selectieve Migratie*, Veenman, Wageningen.

Lukkes, P. (1974) *Regiologie*, Inaugural Lecture, Groningen.

Mas, P. de, Pascon, P. and Wusten, H. van der (1981) *A Balance of Curses and Blessings*, Sociaal-Geografisch Instituut, Paper 1, Universiteit van Amsterdam.

Mik, G. (1980) *Segregatie in Rotterdam*, Final Report, Economisch-Geografisch Instituut, Rotterdam.

Nederlandse Ontwikkelingsgebieden, de (1960) ISONEVO *et al.*, The Hague.

Nota inzake de Ruimtelijke Ordening in Nederland (1960) Staatsuitgeverij, The Hague.

Nota over de Ruimtelijke Ordening in Nederland, Tweede (1966) Staatsdrukkerij, The Hague.

Nota over de Ruimtelijke Ordening, Derde (1975, 1976) Deel 1, *Orienteringsnota*, Deel 2, *Verstedelijkingsnota*, Staatsdrukkerij, The Hague.

Onderzoek Ontwikkelingstendenties Nederlandse Stadscentra (1966) Voorenderzoek Utrecht, Gebruik en Beleving van het Centrum door Bewoners van Utrecht en Zeist, 2 vols., SISWO, Amsterdam.

Ontwikkeling van het Westen des Lands, de (1958) Werkcommissie Westen des Lands, Rijksdienst voor het Nationale Plan, Staatsuitgeverij, The Hague.

Ottens, H.F.L. (1976) *Het Groene Hart binnen de Randstad*, Van Gorcum, Assen.

Paassen, Chr. van (1957) *The Classical Tradition of Geography*, Wolters, Groningen.

—— (1962) 'Geografische Structurering en Oecologisch Complex', *TAG*, *79*, 215-33.

—— (1976) 'Human Geography in Terms of Existential Anthropology', *TESG 67*, 324-41.

Ruimtelijke Ordening en Geografie (1967) Handelingen van het Symposium van het Koninklijk Nederlands Aardrijkskundig Genootschap, Brill, Leiden.

Ruimtelijke Ordening in Nederland (1974) *TAG*, *65*, 4.

SATURA II (1981) *Wonen, Verkeren en Recreeren in Amsterdam*, Interimrapportage 3 (Engelsdorp Gastelaars, R. van, Vijgen, J., Wijs-Mulkens, E. de), Sociaal-Geografisch Instituut, Universiteit van Amsterdam.

SATURA II (1982) *Op Afstand Bekeken*, Interimrapportage 2 (Vidakovic, V., Blaas, H.), Amsterdam.

Schuurmans, F. (1981) *Regionale Arbeidsmarkten*, Monografieen Volks-Telling 1971, No. 16, Staatsdrukkerij, The Hague.

Smidt, M. de (1975) *Bedrijfsstructuur en Arbeidsmarkt in een Ruimtelijk Kader*, Utrecht.

Stad in Nieuwe Vormen, de (1963) *TAG*, *80*, 1.

Steigenga, W. (1955) 'A Comparative Analysis and a Classification of

Netherlands Towns', *TESG, 46*, 105-19.

—— (1960) 'The Urbanization of the Netherlands', *TAG, 77,* 324-31.

–—— (1963) 'Van Sociale Analyse naar Sociaal-Ruimtelijke Constructie', Inaugural Lecture, Amsterdam, *TESG, 54*, 1-9.

—— W. (1964) *Moderne Planologie*, Spectrum, Utrecht/Antwerpen.

STEPRO (1978) Onderzoekreeks *Stedelijke Problematiek West-Nederland*, Nota no. 2, Grondslagen en Programma, Geografisch Instituut, Utrecht.

Takes, Ch. A.P. (1948) *Bevolkingscentra in het Oude en het Nieuwe Land*, Samson, Alphen aan den Rijn.

Tamsma, R. (1966) *De Moshav Ovdiem*, Van Gorcum, Assen.

Timmermans, H.J.P. (1979) *Centrale Plaatsen Theorien en Ruimtelijk Koopgedrag*, Ergon Bedrijven, Eindhoven.

Typologie van Nederlandse Gemeenten naar Urbanisatiegraad (1958, 1964) Centraal Bureau voor de Statistiek, De Haan, Zeist.

Urban Core and Inner City (1967) Proceedings of the International Study Week, Amsterdam, 11-17 September 1966, Brill, Leiden.

Verduin, J.A. (1972) *Bevolking en Bestaan in het Oude Drenthe*, Van Gorcum, Assen.

Verspreiding van de Bevolking in Nederland (1948) IA, Tekst, IB, Afbeeldingen, Publicatie no. 3 van de Rijksdienst voor het Nationale Plan, Staatsdrukkerij, The Hague.

Vooys, A.C. de (1932) *De Trek van de Plattelandsbevolking in Nedderland*, Wolters, Groningen.

—— (1950) *De Ontwikkeling der Sociale Geografie in Nederland*, Inaugural Lecture, Utrecht, Wolters, Groningen.

—— (1951) 'De sterfte in Nederland in het Midden der Eeuw', *TAG, 68*, 233-71.

—— (1959) 'Western Thessaly in Transition', *TAG, 76,* 31-54.

—— (1964) 'Regional Variations in Birth Rate During the Second Half of the 19th Century', *TAG, 81*, 220-32.

Vries Reilingh, H.D. de (1945) *De Volksheogeschool*, Wolters, Groningen.

—— (1961) 'De Sociale Aardrijkskunde als Geesteswetenschap', *TAG, 78,* 112-22.

—— (1964) 'Schaalvergroting en Territoriaal Bindingsprincipe', *TAG, 81,* 405-14.

West-Nederland, Chaotische Planning of Geplande Chaos (1973) Van Gorcum, Assen.

Wenen, Werken en Verkeer in Amsterdam 1880-1980 (1980) Geografisch en Planologisch Instituut, Vrije Universiteit, Amsterdam.

Winsemius, J. (1945/9) *Vestigingstendenzen van de Nederlandse Nijver-heid*, 2 vols., Staatsdrukkerij, The Hague.

Wusten, H. van der (1977) *Iers Verzet tegen de Staatkundige Eenheid der Britse Eilanden 1800-1921*, Sociaal-Geografisch Instituut, Publicatie no. 3, Amsterdam.

——— (1980) 'Dynamics of the Dutch National Socialist Movement 1931-1935' in S.U. Oarsen, B. Hagtvet, J.P. Myklebust (eds.), *Who were the Fascists?* Universitets-forlaget, Oslo/Bergen.

Wissink, G.A. (1962) *American Cities in Perspective*, Van Gorcum, Amsterdam.

Postscript

Attention is drawn to the following, published after this volume went to press:

H. van Ginkel and M. de Smidt, '75 years of Roman geography at Utrecht', special issue of *Tijdschrift voor Economische en Sociale Geografie*, 74 (5), 1983.

11 JAPAN

Keiichi Takeuchi

Pre-war Heritage

The establishment of departments of geography in Japanese universities took place only at the beginning of this century. The geography departments of the Faculty of Literature, Kyoto Imperial University, and of the Faculty of Science, Tokyo Imperial University, were established in 1907 and 1918 respectively. The geography department of the Tokyo University of Science and Literature (Tokyo Bunrika University, which changed its name to Tokyo Kyoiku University after the Second World War and which, still later, transferred to Tsukuba to become the present University of Tsukuba) was founded in 1926. Western geography *per se*, however, had already been introduced into Japan at the beginning of the nineteenth century and, after the establishment of the Meiji government system of education in 1872, geography came to be considered an important course in school curricula. Previous to its institutionalisation at the university level, geography was taught in higher educational institutions such as teachers' training colleges and commercial colleges. Considerable numbers of geographical books were written by professors at these establishments, as well as by certain intellectuals who were interested in geography; many of these works, however, were translations or adaptations of foreign geography textbooks. It should be noted that the above authors wrote these books in the second half of the nineteenth century, naturally for use as texts and reference works, but also as literature aimed at the enlightenment of the Japanese people. Geographical knowledge of industrialised and colonised countries served the mass educational drive towards Westernisation and modernisation, which were almost synonymous in the Japan of that period.

The establishment of geography departments at universities gave rise to the emergence of trained academic geographers who founded study circles and associations for geographical studies and also changed the character of already existing geographical societies. The latter, originally consisting of upper-class dilettantes fascinated by geography, were, consequently, transformed into organisations of research into the earth sciences. Both the newly established and the reorganised geographical societies commenced issuing geographical journals, some of which still continue to exist, one example being the *Chirigaku Hyoron* (*Geographical Review of Japan*), organ of the Association of Japanese Geographers.

Thus, the period of the 1930s was a prosperous one for geographical studies in Japan. Moreover, many books dealing with various branches of geography were published, partly in answer to the demand for them on the part of applicants for secondary school teachers' licenses. During these years, clearly differentiated schools of geography came into being. The school consisting of the staff and graduates of Kyoto Imperial University was characterised by the strong interest in historical geography initiated by T. Ogawa and G. Ishibashi: Ogawa was the first chairman of the geography department there, and Ishibashi the second. The school of geography based at Tokyo Imperial University under the chairmanship of N. Yamasaki and, later, Taro Tsujimura, emphasised physical geographical studies, geomorphology in particular. Tsujimura also stressed the importance of the morphology of the cultural landscape; he was influenced by the German school of *Landschaft* and the Berkeley school. But, unlike these latter schools, Tsujimura and his followers did not attach much importance to genetic studies. The methods of some members of the Tsujimura school tended towards quantitative analyses of distribution, resulting in pioneering works in quantitative geography, although the authors themselves were not aware of the originality of their methods. The characteristics of the school of the Tokyo University of Science and Literature were bestowed on it by K. Tanaka, who established his own unique methodology in defining regions, a methodology somewhat reminiscent of Lauthensach's *Formenwandel*, and by K. Uchida, who laid special stress on regional geography. Apart from adherents of these main schools, several other influential geographers emerged during the same period. Hiroshi Sato specialised in economic geography, introducing it, with particular reference to locational theories, to Japan, and K. Iizuka (1932), after a period of study in France, introduced the Vidalian tradition. R. Ishida insisted on characterising geography as a social science. Subsequently, H. Sato, K. Iizuka and R. Ishida were to exert a strong influence on the younger generations of geographers after the Second World War.

The period of the Second World War was a setback for geographical studies in Japan. Many geographical journals were compelled to cease publication, while others merged to form single, thin issues under the wartime economy. A large number of brilliant young geographers were mobilised and dispatched to the battlefronts, where they died. A certain number of geographers became advocates of so-called Japanese geopolitics, abandoning scientific studies in the interests of the imperialist war for the construction of the 'Greater East Asia Co-prosperity Sphere'. Subsequently, during the post-Second World War period of occupation by the Allied Forces, the leaders of Japanese geopolitics were purged from public posts. The affair involving the rise and eventual fall of Japanese geopolitics was indicative of the weakness of the scientific spirit of the Japanese geographers involved and had a detrimental effect on geography, as a whole, in Japan.

During the wartime, some geographers were sent to the Japanese colonies of that period (Korea, Formosa and Manchuria) or Japanese-occupied areas (China and South-east Asia) to engage in administration and land planning. Few scientific results were obtained by them, however, except for a certain amount of geomorphological studies in Korea and Inner Mongolia, and some ethnographical studies in which a few geographers participated, in co-operation with ethnologists.

Despite differences in study trends among schools of geography, the dominant paradigm in geographical studies before and during the Second World War was the man-environment relationship or environmentalism. Relying on this paradigm, the majority of Japanese geographers had no doubt whatever with regard to the basic unity of geography, except for a few geomorphologists and climatologists who considered their studies to be a part of natural science, i.e. earth sciences. Within the frame of this paradigm, the majority of geographers accepted, as a matter of course, inductive inference based on descriptive methodology. In the 1930s, a few exceptional young geographers, such as I. Matsui (1931) and S. Yoshimura (1933), as well as T. Murata, adopted the quantitative method with deductive inference; but, finding themselves unable to persist in the revolutionary character of their studies in the face of the conventional criticism of the dominant geographers of the time, who had unquestionable faith in the validity of their own inductive methodology, they did not succeed in fully developing the new methodology.

In spite of the wartime stagnation in studies, it was still possible to point out an institutional heritage left over from the pre-war period in the shape of the many scientific organisations, such as the departments of geography in universities, the Land Survey Division of the Imperial Army, meteorological and other experimental stations and geographical societies that were still functioning after the war's end. Formations of geographers arose in Japan, though some of the geographers carried out their postgraduate studies in foreign countries, mainly the United States and Germany. Studies by foreign geographers were fairly well-known among leading Japanese geographers of the time preceding the outbreak of the Pacific war. As for relationships with other disciplines, and also with the indigenous tradition of geography in Japan, Japanese geographers had a fairly close affinity with geology and history, partly because the first generation of Japanese academic geographers was comprised of scholars in one or other of these two disciplines. But they were somewhat isolated from economists, sociologists and researchers in the humanities, including folklore studies, which were attempting a re-evaluation of the indigenous scientific tradition of Japan. While, in this way, we are able to affirm positively the fact of the existence of a school or schools of Japanese geography in the pre-war period, at the same time modern Japanese geography has also to be considered as having the nature of an imported discipline. Debates on the nature of geography, especially on the variety of environmental explanations, were generally made with reference to foreign literature on the subject. Again, apart from the fanatic geopoliticians of the Kyoto school, who insisted that the nature of the Japanese Empire was a Shintoist one and unique, there existed some exceptions to the rule for both scientific isolationism and Eurocentrism. A few professors of geography in commercial colleges, for example, felt compelled to seek a geographical *raison d'être* in the social sciences through daily contact with their colleagues. Tadamichi Odauchi and a small number of other geographers maintained close contact with Japanese folklore scholars, such as Kunio Yanagida, and tried to base their geography on the daily life and legends of the common people of Japan.

Post-war Institutional and Intellectual Conditions

As a result of the administrative reform carried out at the instigation of the Allied Powers occupying Japan after her defeat in the Second World War, many scientific institutions concerned with geographical studies changed their names and official affiliations. The new titles did not, however, effect the functions of these institutions, which remained much the same as before, showing neither signs of expanding nor of being revitalised thereby. The Land Survey Division of the Imperial Army thus became the Geographical Survey Bureau (later it was renamed the Geographical Survey Institute) affiliated with the Ministry of Construction; and the Hydrographical Office was separated from the

Navy to become attached to the Ministry of Transportation.

The administrative reform most important for geographical studies after the Second World War was that involving the education system. This took place in 1949 with the result that former higher educational institutions such as upper secondary schools (comprised of 12th-14th grades inclusive, the first grade being the first year of elementary school) and professional colleges, as well as former national and private universities, were all reorganised to form more than two hundred universities under the new system, which was modelled on that of the American. Many departments of geography were established in these universities; even those universities without departments of geography had at least one chair of geography in the liberal arts course. This resulted in a great increase in the number of professional geographers in universities and also in the number of geographical societies and publications. In pre-war times the Association of Japanese Geographers consisted chiefly of Japanese geographers in the Tokyo area. During the war, however, geographers from other areas began to join the Association; hence, after the war it became a truly nationwide organisation. In 1939, it had 259 members; in 1949, the membership had risen to 549. The number steadily increased to become more than 3,000 today. Other nation-wide geographical organisations, which were founded after the Second World War, are the Human Geographical Society of Japan, the Association of Economic Geography, the Association of Historical Geography, the Association of Professional Geographers and the Japan International Cartographic Association. Considered part of the social studies programme, geography became firmly entrenched in curricula at the junior and senior high school levels; this, consequently, gave rise to a demand for a great number of trained teachers of geography, some of whom became members of geographical societies, in order to dedicate themselves to research activities.

The close of the war put an end to the cultural isolation of the war years and information on new trends in geography in foreign countries became available to Japanese geographers. American influences were strong, particularly in the initial stages of the post-war period, due to visiting Fulbright professors and several American geographers, such as E.A. Ackerman, who worked as consultants to the occupation authorities. After the middle of the 1950s cultural exchanges with European countries also became active, with many Japanese geographers of the new post-war generation pursuing postgraduate studies not only in the United States but also in Europe. In 1957, a regional conference of the International Geographical Union was held in Japan; this constituted an important stimulus to the further development of geographical studies in Japan.

During the pre-war period and throughout the war in Japan, under a regime that was uncompromisingly chauvinist, militarist and nationalist, freedom of speech and thought was greatly curtailed. In geographical publications, for instance, studies of negative aspects of regional policies or resource utilisation were expressly forbidden. Any and every publication dealing with Marxism or materialism was banned; consequently, it was often impossible to publish scientific analyses of regional economy because the censorship authorities would consider studies of this kind to be, methodologically, Marxist-inspired and refuse to allow their publication. The so-called democratisation of Japan promoted by the Allied Powers during the occupation brought about the abolition of this sort of restriction. They did away with the authoritarian character of many geographical associations, an action which together with the liberalisation of studies and study techniques promoted a variety of new forward-looking trends in geographical studies in Japan. This held true especially where the younger generation of geographers was concerned. The latter not only enjoyed the freedom of speech and thought newly bestowed then and which was part of the intellectual climate of post-war Japan, but also proceeded to adopt the analytical methods of neighbouring disciplines, i.e. earth sciences for physical geography and social sciences and the humanities for human and social geography. In the past, authoritarian professors of geography had often prohibited their students from relying on analytical methods such as these, upon the pretext of maintaining the purity and originality of geography.

In the stage immediately after the war, based on the new institutional and intellectual circumstances delineated above, a number of new trends in research appeared, some of which still continue to prosper to this day. Other trends were to appear only to disappear in the face of the ensuing severe criticism that was a form of reaction of one stage to another in the successive stages of post-war change; they would then be replaced by still yet other trends.

One of the most remarkable of the post-war traits was and is the specialisation of each field or branch of geography. For the majority of geographers, physical and human geography are considered to be separate disciplines. Even within physical geography and human geography, respectively, specialisation was accentuated with geomorphology, hydrology and climatology being discrete fields within physical geography and cultural, industrial and agricultural geography being discrete fields within human geography.

New Trends in Physical Geography

In the field of physical geography, as a reaction against the static or landscape analysis-orientated studies, dynamic or genetic analysis came into vogue. Geomorphological studies were very often made under the title of landform development, and landform developments in lowlands were new subjects for intensive study. The development of alluvial plains and coastal lowlands, especially on the Pacific Coast, were first summarised in the early 1950s by A. Sugimura (1950), F. Tada *et al.* (1952), T. Yoshikawa (1953), T. Nakano (1954) and S. Kaizuka (1955). They were more or less influenced by pre-war studies on Quaternary history in Japan by geologists such as Y. Otsuka (1931); but they have refined these earlier works with new findings on sub-aerial and submarine topography. Studies pertaining to these topics progressed greatly with the introduction of tephrochronological methods in the analysis of terrace topography (Kaizuka, 1958). Moreover, the coastal development of Japan came to be discussed in relation to glacio-eustatic changes in sea level in the Quaternary (Yoshikawa, Kaizuka and Ota, 1964).

In the analysis of fluvial processes and of the formation of coastal landforms, quantitative approaches were promoted by Y. Mino (1950), E. Yatsu (1955) and others. Furthermore, in the studies of the landform development of mountain regions, Japanese geomorphologists began to consider the development of river terraces in relation to mountain glaciation, the late Quaternary eustatic changes in sea level and other phenomena related to Quaternary climatic changes. K. Kobayashi (1958), T. Yoshikawa (1953), S. Fukai and others discussed the relationship between tectonic and climatic factors in the development of river terraces. Many drainage basins of larger rivers in central and northern parts of Japan were studied with special reference to the evolutionary relationship between mountains and their adjacent basins and lowlands; geomorphological studies have thus evinced steady progress, resulting in synthetic works such as, for example, Y. Sakaguchi's morphometric study of the Japanese islands (1980).

In the field of hydrology, dynamic and quantitative approaches had already been introduced in the pre-war period; examples of these are the pioneering works of S. Yoshimura (1942). Since the war, remarkable progress has been made with the introduction of new research techniques and also under the stimulus resulting from the application of these techniques to resource development, agriculture and so on. Leaders in this field of study were S. Yamamoto and M. Ichikawa and the researches initiated by them (Yamamoto, 1953, 1968, 1972; Ichi-kawa, 1973) have enriched the studies of the younger generation of geographers.

Prior to the war, pioneering works in climatology by meteorologists and geographers had already appeared, notably those of E. Fukui (1938) and T. Okada. The interest in climatology on the part of geo-graphers was not very pronounced during the stage immediately after the war, except for T. Sekiguti (1951) and T. Yazawa (1953, 1956), whose pioneering works on local climatology, synoptic climatology and climatic landscape appeared at that time. After the middle of the 1950s, studies of climatology were vigorously promoted by geographers of the younger generation (I. Kayane, 1966). Climatological studies still cover a wide range of topics. Particularly remarkable work has been accomplished in the fields of urban climatology (Nishizawa, Yamashita and Suzuki, 1979), meso- and micro-climatology and large-scale dynamic and synoptic climatology. Studies of urban climate have un-failingly included the viewpoint of environmental or applied studies such as those on pollution problems. Studies of large-scale climatology are characterised by the introduction of dynamic or geographical approaches and also by the world-wide viewpoint. M.M. Yoshino main-tains the leading position in many areas of climatological studies in Japan (1957, 1961, 1971, 1975); T. Nishizawa has made numerous con-tributions to the study of urban climate; and H. Suzuki has accom-plished remarkable work in the fields of dynamic climatology and his-torical climatology (1962, 1975).

Here we are not discussing in detail the new trends in other fields of physical geography such as biogeography, for example. There are cer-tainly a number of works worthy of consideration in these other fields — M. Momiyama's series of works on medical geography, to men-tion only one example (1977). However, the number of geographers who are specialised in biogeography and medical geography is rather limited and, moreover, many of their works have been made in an en-vironmentalist context and should not be considered solely in the framework of physical geography. M. Momiyama herself has insisted on the necessity of the social science viewpoint in medical geograph-ical studies (1947, 1949, 1950).

The geographical bibliography, edited by the Human Geographical Society of Japan since 1947, shows that until the middle of the 1950s papers on physical geography made up less than 30 per cent of all geo-graphical publications. Since then, however, the percentage of physical geography in all publications has been steadily increasing to reach, at this time, more than 40 per cent: new approaches initiated in each field after the Second World War have made generally consistent and steady progress in the course of the past 40 years.

New Trends in Human Geography

Environment and People

The situation in human geography is rather different from that in physical geography; for one thing, with the former there have been trial and error or paradigm shifts in the approaches. The first notable aspect of new trends in post-war human geography was that, in the period immediately after the war, the methodology and the philosophy which had heretofore characterised it and which had been widely accepted came in for some severe criticism. The predominant tendency in the immediate post-war period of laying stress on socio-economic and historical causalities in geographical interpretations was, in part, a reaction to the perverted and superficial landscape studies of pre-war times. It derived also from the criticism directed towards environmentalism. It should be pointed out, though, that this criticism was somewhat exaggerated or even unfounded. For, while some geographers fervently continued to attack the supposed enemy that was environmental determinism, the latter was, by then, in fact obsolete. Discussions involving this subject were very often carried out on a metaphysical level. Those who were critical of the subject, however, subscribed either to the so-called possibilist viewpoint that shared the same ecological basis – that of modern classical geography – with environmentalism, or to the more or less dogmatic Marxist viewpoint that declares the role of the natural environment in the development of society to be merely 'important but not decisive'. These arguments were not only unproductive but, to a certain degree, they also hindered geographical studies from attaining a true ecological perspective on the assessment of the environment. In this sense, these arguments contributed to a stagnation in the stance or thinking proper to geography, among some circles of geographers.

The emphasis on socio-economic and historical causalities assuredly encouraged intensive empirical studies of small spatial units. This was true in every field of human geography – economic, social and settlement geography. This research method of giving importance to field observation had been a tradition in human geography in Japan since pre-war times, but there remained a gap between intensive local studies, on the one hand, and theoretical and nation-wide synthesis, or comparative geographical studies, on the other.

Historical Geography

The second notable aspect of immediate post-war trends in human geo-graphy was the flourishing state of historical geographical studies. As mentioned previously, the study of historical geography is firmly rooted in the school of the geographical department of Kyoto University. Besides the adherents of this school there was K. Uchida in the geographical department of Tokyo Bunrika University, who insisted that, while the aim of historical geography should consist of the elucidation of various present-day geographical conditions, it was most important to examine the historical character of these conditions. Uchida continued to work in the post-war period (1971) and the academic tradition which he established and which focuses on the seventeenth and eighteenth centuries (early modern period) has been succeeded to by his followers Y. Asaka, T. Kitamura and T. Kikuchi (1958).

Historical geography was understood rather differently by the Kyoto school. S. Komaki, third chairman of the geographical department of Kyoto University, had, prior to the war, already systematised their standpoint (1937). According to Komaki, the work of historical geo-graphy was the restoration of the landscape of past times. After the war, Komaki retired from the forefront of academic circles because of his involvement in the affairs of 'Japanese geopolitics'. His method-ology in historical geography has been faithfully upheld by his fol-lowers, K. Fujioka (1955-77, 1977), J. Yonekura (1949) and others, who have exercised remarkable leadership in the development of studies of historical geography in post-war Japan. Immediately after the Second World War, their interests tended to focus on the prehistoric and ancient periods. The most noteworthy contribution by historical geographers in this field has been the reconstruction of the *jori* land-division system in various parts of Japan (T. Tanioka, 1959; H. Wata-nabe, 1968). In the 1950s, the academic tradition of the Kyoto school was applied by historical geographers of the younger generation to the studies of the mediaeval and early modern periods.

The differences between the Kyoto and Tokyo schools of historical geography became blurred with the passing of time, mainly due to the pervasive influence of the Kyoto school, on the one hand, and to the general recognition of the importance of historical causalities, on the other. Instead of the 'reconstruction of past landscapes', the term 'reconstruction of spatial organisations of past times' is now used by, for instance, T. Tanioka. Some historical geographers have become interested not only in the reconstruction of past landscapes, but also in the perception of the people of past times, as shown in the studies of M. Senda (1980).

Urban Studies

Many new trends in human geography were the reflection of new aspects of early post-war Japanese society. The rapid reconstruction and growth of the Japanese economy gave rise to various social problems, together with significant changes in the distribution of the nation's population. Previous to the war, urban studies had not been popular among Japanese geographers, except for S. Kiuchi with his pioneering works (1951), and a very few others. After the war, there was a remarkable increase in interest in urban geography; during the first years of the post-war period, a number of urban geographers investigated many cities damaged by the war and the analytical recording of the reconstruction of these cities was welcomed by geographers of both academic and practical persuasions. Industrialisation and urbanisation after the 1950s helped direct geographers' interest towards urban areas (Kiuchi, Yamaga, Shimizu and Inanaga, eds., 1964). Apropos of this, some academic geographers were mobilised into an involvement on practical concerns such as town planning.

A large number of the studies pertaining to these matters were empirical or local in character, but it was also possible to observe in them the introduction of new methods in urban research, originally developed mainly in Anglo-Saxon countries. New approaches in urban geographical works in the 1950s are observable in the studies of commuting and urban traffic (Arisu, 1968), of the internal structure of metropolitan areas (Tanabe, 1971), of satellite cities (Yamaga, 1964), of urban influence in surrounding rural areas, of the hierarchy in urban settlements (Y. Watanabe, 1955), of the quantitative analysis of rural-urban relationships and shopping districts (Hattori and Sugimura, 1977), and so on. Major contributions to the progress of urban studies were made by the research group on urbanisation of the Association of Japanese Geographers. In the 1950s and 1960s the Association of Japanese Geographers organised symposia and research meetings which were very instrumental in increasing the interest in urban geography.

The emergence of urban problems in the immediate post-war years and the subsequent period of rapid economic growth stimulated not only urban geographical studies but also urban studies in many of the social sciences and the humanities. The progress of urban studies in adjacent sciences has, in turn, naturally stimulated further studies in urban geography, creating increased opportunities for co-operative or joint research by specialists in various fields. This tendency has become further pronounced since the end of the 1960s with the increased concern over civic problems in urban areas, such as environmental pollution, traffic congestion, increased prices in housing and difficulty in the realisation of proper zoning. Until the middle of the 1960s geographers and sociologists generally took the initiative in co-operative research into urban areas. With the emergence of many problems relevant to urbanisation, the further participation of specialists in economics, political science and jurisprudence in co-operative research has become imperative. Very often, the latter groups of specialists now take the initiative as leaders in joint research. Under these circumstances, as with other fields of human geography, geographers have found themselves compelled to discover and define their scientific identity or, otherwise, to convince specialists of other disciplines of their proper competence.

Economic Geography

The emergence of new socio-economic aspects in post-war Japan, which have stimulated human geographical studies, has not been limited to the field of urban geography, for the changing character of manufacturing industries has greatly stimulated new studies in industrial geography. In pre-war times there existed a rift between theoretical or locational studies and regional or empirical studies. In the 1950s historical geographical approaches and the studies on rural or local industries tended to predominate. Later, however, studies in manufacturing geography were actively produced, reflecting the rapid economic growth in Japan from the late 1950s through the 1960s. Many of these were analyses of industrial areas in the 1960s (A. Takeuchi, 1973), while the number of studies on nation-wide industrial location increased in the 1970s. In rather sharp contrast with traditional descriptive analyses of industrial areas, the analysis of spatial organisation concerning manufacturing was initiated only in the 1970s (Kitamura and Yada, eds., 1977). Tracing the flow of capital and goods in the framework of the total manufacturing process is somewhat difficult, especially where the complicated subcontract system forms the bulk of the support for the manufacturing process, on the one hand, and where big, oligopolistic enterprises manage the spatial flow of capital and goods and also the spatial division of market areas, on the other (Kawashima, 1976). These emerging studies on manufacturing geography have been carried out mainly by the post-war generation of geographers and have contributed greatly to an understanding of the radical changes in the spatial organisation of Japanese industries, during both the period of rapid economic growth in the 1960s and the low growth rates since 1974.

Quantitative analyses are relatively few in number in the field of manufacturing geography, in comparison with the number in the field of settlement geography and the geography of tertiary activities. Moreover, agricultural and rural geography have shown great changes in topics of study in the past 40 years since the war, reflecting the changing character of agriculture and rural communities throughout these years. A small number of theoretical or local works by geographers, such as those of S. Aoga (1940), and many empirical and local studies, which were descriptive in character, already existed before the war. In the period immediately after 1945, some geographers who had previously been engaged in agricultural geography published works summarising their long years of study in this area (Sugai, 1949; Kagose, 1953). In these works, we are also able to observe the reflection of the economic situation of that stage, in which the increase of agricultural production was the prime and urgent task.

Since the Second World War, however, locational analyses in agriculture have been carried out, with few exceptions, by specialists in agricultural economics. Until the end of the 1950s the main targets of agrarian policies in Japan were the increase of staple food production and the protection of the small peasantry economy. Problems of reclamation and land improvement were rather popular themes of geographical studies and the changing patterns of distribution of different agricultural products (Ebato, 1969; S. Yamamoto, 1972) was and continues to be analysed by large numbers of geographers. Moreover, since the beginning of the 1960s, Japanese agriculture has been influenced by urbanisation and industrialisation, and huge rural and agricultural exoduses have subsequently occurred. In addition to these factors, the liberalisation of foreign trade and agricultural overproduction have caused substantial changes in agrarian politics, from the protection of the small peasantry economy to structural reform for the purpose of the establishment of entrepreneurship in Japanese agriculture. Since the middle of the 1960s, geographers have thus directed their attention towards the impact of industrialisation and urbanisation on agriculture and farm villages. They have paid special attention to problems such as the increased percentage of part-time farm households (Ishii, 1980), the formation of new specialised crop areas, and the phenomena caused by the massive exodus of manpower from the remoter marginal areas – mountains, isolated islands, and so on.

Agricultural and rural geography have certainly registered remark-able progress in studies in the post-war period. If we take into consider-ation the tradition of studies from the pre-war period, the relevance of problems pertaining to agriculture and rural communities, the increased number of available researchers and the increased facilities for study such as statistical data and the utilisation of computers, we may very well observe that all this progress is merely a matter of course. How-ever, there are certain problems here that are common, in some respects, to other fields of human geography, but which appear more explicit in this particular field of geographical study; I refer to the problems per-taining to the capability and competence of the geographers them-selves, and to the question of value commitment in their researches.

A large number of agricultural economists, rural sociologists, his-torians and other specialists in various other disciplines are engaged in studies of agriculture and rural communities, as are researchers and technicians of ministries and other governmental organisations. Very often the latter can avail themselves of more useful material and data than academic researchers can. Geographers specialised in the agricul-tural field must, perforce, demonstrate their *raison d'être* in competi-tion with two groups — academic researchers in neighbouring discip-lines, on the one hand, and research members in technology or bureau-cracy, on the other.

The second problem facing researchers is almost inevitable in the analysis of present-day agriculture and rural communities. The ques-tions of whether priority should be given to the efficiency of the national economy, or whether the inhabitants of depopulated villages in marginal areas have the right to continue to live there with the support of huge government subsidies (Fujita, 1981) and so on, impose the necessity of a value or policy commitment upon the geo-graphers concerned, who must help formulate answers to these ques-tions. Not very many geographers have dared to accept this value judge-ment, which has certainly decreased the influence of geographers in practical concerns.

With the increasing number of academic geographers in Japan, many fields hitherto studied by only a few have become important topics of study for a large number of geographers. Although mineral produc-tion, including coal mining, had decreased in importance in the national economy, some remarkable studies in this sector were accomplished in the immediate post-war period. Before the war, Y. Yamaguchi had pub-lished numerous studies on coal-mining settlements, and his approach in studying mineral production in the framework of settlement geography has been adopted by a certain number of geographers such as S. Kawasaki (1973). Moreover, studies of the distribution network of mineral products and of changing features of coal-mining areas, along with the delay of the coal-mining industry and coal-mining disasters, have also been published (Yada, 1975).

Marine production was studied by Japanese geographers before the Second World War mainly as the geography of fishing villages. A leading role in this field was played by H. Aono (1953). After the war the interests of geographers were focused not only on fishing villages but also on the distribution system, problems concerning the cultivation of pearls, edible seaweeds, shellfish, certain kinds of fish (Oshima, 1972) and on the symbiotic relationship between fishermen and farmers (Birukawa and Yamamoto, eds., 1978). Reflecting the decreased share of coastal fishery in the marine production of Japan after the Second World War, there appeared a growing number of studies on major fishing ports, which served as bases of offshore and deep-sea fishery.

Prior to the war population geography had been one of the favourite study themes of Japanese geographers. Their specific interests in this field were, however, distribution and population potentials in the frame of an environmental analysis. Since the war's end, with the increased availability of statistical data and the progress of analytical techniques, more detailed and sophisticated studies have been produced. In this field, however, it is admittedly somewhat difficult to distinguish geographical works clearly from those written from the viewpoint of demography. It was in the capacity of geographer, for example, that H. Hama published numerous works which were highly estimated by demographers (1978).

Geographical studies of distribution (transportation, communication and commercial activities) were effectively initiated in the postwar period. Before the war, only a limited number of historico-geographical studies had been made on transportation; at the same time, some attempt had been made at delimiting the hinterland area. After 1945 a considerable number of intensive regional studies incorporating data on transportation, with particular reference to commuters and passengers, were written by geographers such as T. Arisue (1968), K. Sawada and K. Shimizu. Only after 1960 did the quantitative approach in studies in the above field begin to appear. In the 1950s, in commercial geography, Y. Sugimura had initiated empirical researches into the shopping centres of various Japanese cities. In the following decade H. Morikawa (1974) and others adopted the quantitative approach, which they commenced to apply to the central-place pattern in Japan.

Where the field of tourism and recreation was concerned, in the pre-war period only a few sporadic studies on recreational spots, notably spas, were published. In the early post-war period, too, the main interests of geographers researching recreation areas were centred on spas, which are numerous in Japan and frequented by many tourists. It was only after the middle of the 1960s that geographical studies of other types of tourism activities and their development began to emerge. This new trend reflected the emergence of new patterns in the leisure behaviour of the Japanese.

Other Areas

In the field of political geography, there was a disruption in the production of studies in the period immediately after the war. This constituted a reaction against geopolitics, which had had so great a vogue during the war; but it did not necessarily signify theoretical reflections on, or criticism of, Japanese geopolitics. After the 1960s various studies classified under the heading of political geography appeared. Most of them were either studies of certain administrative districts, or historical (Hayashi, 1961; Iwata, 1962), or rather sociological, studies of the organisation of the village community. No study has yet been made from the viewpoint of the analysis of the spatial aspects of power.

In the area of cartography, most of the notable works produced in the early post-war period were formulated by governmental institutions, especially the Geographical Survey Institute and the Hydrographic Department. It is impossible to enumerate here a complete list of all the thematic maps and charts produced by these institutions. The National Atlas of Japan, published in 1977, synthesises these works. Besides some works on map projection, many studies have been carried out on map history, constituting the academic contributions by geographers to cartography. In this field, not only Japanese old maps, including pictorial specimens, but also Chinese old maps and Western old maps of Japan and other parts of the Far East, have been subjected to scrutiny by numerous Japanese geographers. T. Akioka (1971) and N. Ayuzawa, who had already published works prior to the war, were pioneers in this field. Since the war's end a large number of geographers, such as K. Unno and K. Yamori, have contributed to the further development of these studies. Their studies are formulated not only in order to present the viewpoint of historical geography but also with the purpose of exploring and analysing the characteristics of the cosmology and the spatial perceptions of the Japanese and other Asian peoples

(Kyotodaigaku Bungakubu Chirigaku Kyoshitsu, 1982).

Most of the empirical or field studies of post-war Japan have been conducted in Japan. As mentioned before, in the pre-war period, when Japan possessed colonies or wielded influence over certain overseas territories, her geographers, nonetheless, did not succeed in producing many remarkable studies in fields outside Japan. It was only after the middle of the 1950s that some geographical studies in the field in foreign countries began to be carried out. Some were accomplished by individuals who had opportunities for studying abroad with the aid of grants from Japanese or foreign institutions, others by members of research project teams subsidised by grants from the Ministry of Education. In the following decades there at last developed a number of geographers who specialised in defined areas or countries of the world at large.

Schools and Scholastic Groups

Debates on the nature and methodology of geography occurred in pre-war Japan. In the fields of physical geography, the problems discussed then, however, have not had any measurable impact on post-war period discussions. Since the war, specialists in physical geography have always harboured − either implicitly or explicitly − the common concept that physical geography was a natural science, constituting a part of the earth sciences. The so-called methodological discussions in physical geography after 1945 rather concerned research techniques. Thus, in physical geography after the Second World War, there are no distinct schools in the pre-war sense, such as the school of the University of Tokyo or the school of Tokyo Bunrika University. Various scholastic groups have indeed been formed, but this has been done according to interests and research techniques; for example, there might be found a group of geomorphologists interested in the development of plains and continental shelves; or a group engaged in the study of glacial topographies or in urban climate studies, and so on. These research groups, moreover, generally have an inter-university character.

In human geography, the situation has been rather different. Methodological discussions that took place in the pre-war period were resumed in the post-war era, and geographers who had been involved in the pre-war discussions exercised a great deal of influence on the latter. Furthermore, new methodological debates took place after the Second World War. Some involved geographers from different universities or schools of Japanese geography; but many others were carried on in a framework reflecting the institutional and academic characteristics of one or other of the schools.

R. Ishida, who was a graduate of the University of Tokyo, insisted on the importance of the viewpoint of the social sciences for geography in the 1930s, and criticised the application of natural science methods such as that involving a gravity model, for example, to human and social phenomena (1933). He has a large number of followers among the post-war generation of geographers, from a wide range of universities throughout the country. He was assuredly responsible for the encouragement of some new scientific trends in post-war geography, which lay stress on the importance of historical causation, on the human and social nature of the geographical environment, and on the necessity of deliberately considered criticism of source materials. But, on the other hand, with all his influence he without doubt impeded to a certain degree the advance of a quantitative revolution in Japan.

K. Iizuka was another influential geographer, at least immediately after the war, especially among younger geographers of the school of the University of Tokyo. He had already insisted on the importance of socio-economic factors in geographical analysis in his works on the history of modern geography over a period of some years before 1945. In several other books, published in the decade after the war, he continued to assert his position (1947). He has never explicitly adopted a Marxist stance, but many young geographers of the early post-war period, under his influence, admitted their viewpoint to be a Marxist one, and one which at times could even be considered somewhat dogmatic, in line with the predominant Stalinism of the early 1950s. As mentioned before, this Marxism-orientated group ignored efforts to discover a methodology proper to geography but contributed to the development of research in economic geography expressed in terms of the economic sciences. One further fact to be pointed out regarding this group is that it has not, however, succeeded in synthesising its anti-environmentalistic stance, its Marxist theoretical viewpoint and the empirical field studies which it has practised and in which emphasis has been placed on historical causalities.

In the University of Tokyo, apart from the department of geography of the Faculty of Science, a new department of geography was established, in 1957, in the Faculty of Liberal Arts, under the chairmanship of S. Kiuchi. Though some of the staff and graduates of this department have been strongly influenced by the Iizuka school, they claim, to a considerable extent, to have rediscovered the original character or *raison d'être* of geography. In a certain sense, they are faithful to the traditional concepts of geography and have, as well, a considerable awareness of the achievements of orthodoxy in geography in foreign countries. They contributed a great deal to empirical studies, especially in the fields of settlement and applied geography.

The situation in the geographical school of Tokyo Kyoiku University has been somewhat different. Many graduates of this university have occupied the chairs of geography in numerous local universities and have been actively engaged in intensive regional studies. They are open-minded regarding the adoption of new research techniques and new interests developed by foreign geographers. A large number of the younger geographers of this school have pursued postgraduate courses in foreign countries and have also participated in scientific expeditions abroad. Moreover, the quantitative method has been widely adopted by this school and its adherents have consistently explored new fields of geography such as mental maps, the spatial organisation of financial activities, and so on. We may very well say that they have formed a new orthodoxy within the academic geography of Japan.

As previously pointed out, the Kyoto school is generally committed to historical geography. The fact that there are today a number of cultural anthropologists and ethnologists, who were graduates of geography departments considered to belong to the Kyoto school of geography, is worthy of note. In the sense that cultural anthropology or ethnology are essentially anti-historicist, this may seem somewhat curious; but it can be explained by the great interest felt by this school in the neighbouring disciplines of the humanities in the post-war period. We should also note this school's strong interest in cultural geography, an interest which was also a tradition, fostered by means of the historical geographical studies special to this school. While the morphology of the cultural landscape has been almost completely discarded as a subject of research by the school of the University of Tokyo, some of the adherents of the Kyoto school – for instance, I. Suizu (1969) – have undertaken morphogenetic studies in cultural geography. Thus they have succeeded to the best part of the heritage of the historical geography of the school of the pre-war period, resulting in the flourishing state of development in this field of study.

At the same time, the adherents of the Kyoto school have given themselves over to methodological reflections in studies pertaining to the history of geography and cartography. Furthermore, these studies include not only the development of Western geography but also the indigenous tradition of oriental geography, especially that of Japan.

Their achievements are summarised in a recent publication, edited by the department of geography of Kyoto University (Kyotodaigaku Bungakubu Chirigaku Kyoshitsu, 1982). The interest of the geographers of this school covers a wide range of fields in human geography; nevertheless, the character of the school as a whole remains very clear and consistent in its interest in the spatial aspects of human activities in every place and in every historical phase, including the present, and in the methodological reflections achieved through the study of the history of geographical thought.

Some departments of geography established after the war are now considered to be independent schools; these include the geographical schools of Tohoku (Sendai) and that of Hiroshima. The characters of these schools are clearly defined by the study themes and the methodological trends to which they subscribe. The latter themes and trends, however, in great part depend on the trends of study pursued by the staff of the geography departments concerned, and thus do not constitute a tradition in the true sense.

Summary

As indicated above, the existence of distinct scholastic groups and schools in present-day Japanese geography is an established fact. We are fully aware that the above description is somewhat schematic and simplified; but, with the current increased facilities of communication and occasions of scientific intercourse, it is very often rather difficult to ascertain the fact of 'belonging-ness' to a school, when an individual geographer is at issue. Since the end of the Second World War contacts with foreign geography have remarkably increased, in every way. Consequently, under circumstances such as these, it is also difficult to make a comparison of the characteristics of Japanese geography with the aspects of geographical studies in other countries. Still, there are some aspects proper to Japanese geography that can be pointed out.

The overwhelming importance of the national territory as a subject of geographical research may not be proper to Japanese geography alone. In every country, after all, it is natural that most parts of geographical studies are focused on the areas of the researchers' own country. In the case of Japan, however, the cultural and geographical isolation, as well as the language barrier, further fosters this tendency towards extensive research on home territory. In addition, most of the results of Japanese researches are published in Japanese, though sometimes accompanied by English summaries. For these reasons, the achievements of Japanese geographers are comparatively unknown outside Japan, in adverse proportion to the large number of existing Japanese geographers and the quantity of Japanese geographical publications.

The weight of interest which Japanese geographers give to different topics of geography displays certain peculiar characteristics. Most of the research in physical geography is dedicated to geomorphology and climatology; in human geography, interest in economic geography – even in studies pertaining to historical or settlement geography – is very strong. Cultural and social (in the sense of 'sociological') interests are relatively weak.

Institutionally, geography is firmly established in the school curriculum and in the institutions of higher education and research. Thus geography is armed with an institutional safeguard, which could give rise to a controversial situation; geographers could, subsequently, engage in vanguard or revolutionary attempts to change the *status quo* in both the methodological and social sense, on the one hand, or they could maintain themselves easily and peacefully in a conventional or conservative position, on the other. In actual fact, many Japanese geographers have adopted the latter stance. In spite of strong tendencies towards a Marxist economic geography, a creative formulation of Marxist theory in geography has not yet been accomplished. Numerous geographers have adopted quantitative geography, but few have constructed original, theoretical models. There are also many who have opposed a quantitative revolution but, again, only a small number have criticised the 'new geography' on the basis of epistemological reflection.

In contrast to the firm establishment of geography in schools and academies in Japan, geography has not yet obtained a strong position in the fields of application such as planning and marketing. There are hundreds of graduates in geography every year but, except for a few who obtain teaching posts or make an academic career of geography, the majority obtain jobs that have nothing whatsoever to do with geography.

The development and prosperity of a discipline in one country must not be evaluated by the number of scholars or the quantity of publications dealing with it, but by the quality of its achievements. Remarkable achievements certainly exist among geographical studies of the past 40 years, which deserve to be known by the international geographical community. The 24th International Geographical Congress held in Japan in 1980 was an opportune occasion for foreign scholars to learn of recent achievements of Japanese geographers and, also, for Japanese geographers to receive a stimulus for their further creative work.

References

There exist several writings in Western languages on modern geography in Japan (Association of Japanese Geographers, ed., 1966; Kiuchi, ed., 1976; Kornhauser, 1979; Pinchemel, 1980; Science Council of Japan, ed., (1980). In ome of them, the bibliographical annotations are quite comprehensive. Furthermore, I have already analysed various aspects of Japanese geography (Takeuchi, 1974, 1978, 1980). Here, I have not aimed at presenting an exhaustive or comprehensive bibliography, but have tried to outline the development of geographical studies, especially those carried out in the post-war period, as a history of thought, taking into consideration the social background of this development.

Akioka, T. (1955) *Nihon chizushi* (History of Japanese Maps), Kawade-shobo.

—— (1971) *Nihon kochizu shusei* (A Collection of Old Maps of Japan; A History of the Making of Japanese Maps), Kashimakenkyusho.

Ando, M. (1964) *Nihon no kaju* (Fruit in Japan), Kokonshoin, Tokyo.

Aoga, S. (1940) *Nogyokeizaichirigaku* (Agricultural Economic Geography), Sobunkaku, Tokyo.

Aono, H. (1953) *Gyoson suisan chirigaku dai-ishu, dai-nishu* (Studies in the Geography of Fishing Villages), 2 vols., Kokonshoin.

–—— (1953) *Gyoson suisan chirigaku kenkyu* (Geography of Fishery and Fishing Villages), 2 vols., Kokonshoin.

Arisue, T. (1968) *Nihon no kotsu* (The Transportation of Japan), Kokonshoin.

Association of Japanese Geographers (ed.) (1966) *Japanese Geography, Its Recent Trends*, Special Publication No. 1.

Birukawa, S. and Yamamoto, S. (eds.) (1978) *Engan shuraku no seitai* (Ecology of Coastal Villages), Ninomiyashoten.

Ebato, A. (1965) *Nihon nogyo no chiiki bunseki* (Regional Analysis of Japanese Agriculture), Kokonshoin.

—— (1969) *Sanshigyo chiiki no keizai-chirigakuteki kenkyu* (Economic Geography Study of Silk-reeling Industry Areas), Kokonshoin.

Ezawa, J. (1965) Morphology and Economic Theory of the Industrial Agglomeration, *Papers and Proceedings of the 1st East Conference Association*, University of Tokyo Press.

Fujioka, K. (ed.) (1955-77) *Nihon rekishichiri sosetsu* (General Remarks on Japanese Historical Geography), 5 vols., Yoshikawakobunkan.

—— (1977) *Gendai toshi no rekishi-chirigakukeki kenkyu* (Historical-Geographical Analysis of Modern Cities), Kokonshoin.

Fujita, F. (1981) *Nihon no sanson* (Mountain Villages in Japan), Chijinshobo.

Fukui, E. (1938) *Kikogaku* (Climatology), Kokonshoin.

—— (1963) *Kikogakugairon* (Introduction to Climatology), Asakurashoten.

Funakoshi, A. (1976) *Hoppozu no rekishi* (History of Maps of the Northern Region), Kodansha.

Hama, H. (1978) *Jinko mondai no jidai* (The Era of Population Problems), NHK Books.

Hamatani, M. (1976) *Gendai noson no chiiki chitsujo to sono henkyo-sasazeki suiri chiiki o jirei ni shite* (The Regional Order of the Modern Village in Transition — A Case Study of the Sasazeki Water Supply Area), *Shirin*, 59-2, 63-97 (227-61), Shigakukenkyukai.

Hattori, K. and Sugimura, N. (1977) *Shotengai to shogyochiiki* (Shopping Streets and Commercial Districts), Taimeido

Hayashi, M. (1961) *Shi cho son no seiji* (Politics of Local Government; Shi, Cho and Son), Kokonshoin.

Ichikawa, M. (1958) 'On the Debris Supply from Mountain Slopes and Its Relation to River Bed Deposition', *Sci. Rep. Tokyo Kyoiku Daigaku, Sec. 6*, 6, 1-29.

—— (1973) *Suimongaku no kiso* (Foundation of Hydrology), Kokonshoin.

Ichikawa, T. (1966) *Koreichi no chirigaku* (A Geography of High and Cool Regions), Reibunsha.

Ide, S., Takeuchi, A. and Sawada, H. (1977) *Keizai chiiki no kenkyu — keihin chiiki no kogyo to nogyo* (Studies on Economic Regions: Industry and Agriculture in the Tokyo-Yokohama Area), Hakubunsha.

Iizuka, K. (1932) *Shakaichirigaku no doko* (Trends in Social Geography), Tokoshoin.

—— (1947) *Chirigaku hihan* (Critics of Geography).

Ishida, R. (1933) 'Chirigaku ni okeru hosokusei — Shotoshite kankyosetsu to ketteisetsu ni tsuite (Laws in Geography — in relation with environmentalism and determinism), *Chirikyoiku, 17*.

—— (1952) *Jinmon chirigaku kenkyugo nyumon* (Introduction to the Studies of Human Geography).

Ishii, M. (1980) 'Regional Trends in the Changing Agrarian Structure of Postwar Japan' in the Association of Japanese Geographers (ed.), *Geography of Japan*, Teikokushoin, 199-222.

Ishimizu, T. and Ishihara, H. (1980) 'The Distribution and Movement of the Population in Japan's Three Major Metropolitan Areas', in The Association of Japanese Geographers (ed.), *Geography of Japan*, Teikokushoin, 347-78.

Itakura, K. (ed.) (1978) *Jibasangyo nomachi jo, ge* (Local Industrial Towns), 2 vols., Kokonshoin.

Itoh, T., Naito, H. and Yamaguchi, F. (eds.) (1979) *Jinko ryudo no chiiki kozo* (Geographical Structures of Population Movement), Taimeido.

Iwata, K. (1962) *Sekiato to hankai* (Sites of Check Gates and Boundary of Clan Territory in Historical Days), Azekurashobo.

James, P.E. (1972) *All Possible Worlds, A History of Geographical Ideas*, The Odyssey Press, Indianapolis, 342-6.

Kagose, Y. (1953) *Kochi no chiriteki kenkyu – sono kaihatsu to kairyo* (Geographical Study on Arable Land), Sundai Shuppansha.

Kaizuka, S. (1955) 'Kanto-nangan no rikudana keisei-jidai no kansura ichi kosatsu' (On the Age of Submarine Shelves of Southern Kanto), *Chirigaku Hyoron*, 15-26.

—— (1958) Tephrochronological Studies in Japan, *Erdkunde*, 254-70.

Kawabe, H. (1980) 'Internal Migration and the Population Distribution in Japan' in The Association of Japanese Geographers (ed.), *Geography of Japan*, Teikokushoin, 379-89.

Kawasaki, S. (1973) *Nihon no kozan shuraku* (The Geographical Studies of Japanese Mining Settlements), Taimeido.

Kawashima, T. (1976) 'Chiiki keisaku kara mita sangyo kozo no tenkan' (The Regional Policy View of Change in Industrial Structure) in *Sangyo kozo tenkan no shojoken* (Conditions of Changes Industrial Structure), Osaka shiritsu Daigaku Keizai Gakkai, 117-42.

Kayane, I. (1966) 'Meso-climatological Research on the Temperature Distribution in the Kanto Plain, Japan', *Sci. Rep. Tokyo Kyoiku Daigaku, 87*, 125-87.

Kikuchi, T. (1958) *Shinden kaihatsu* (Reclamation) 2 vols., Kokonshoin.

Kishimoto, M. (1968) *Nihon no jinko-shuseki* (Regional Concentration of Population in Japan), Kokonshoin.

—— (1978) *Jinko idoron* (The Study of Population Movement), Kokonshoin.

Kitagawa, K. (1976) *Koiki chushinchi no kenkyu* (A Study of Regional Capital Cities), Taimeido.

Kitamura, Y. and Yada, T. (eds.) (1977) *Nihon kogyo no chiiki kozo* (Regional Structure of Japanese Industry), Taimedo.

Kiuchi, S. (1951) *Toshichirigaku kenkyu* (Studies of Urban Geography), Kokonshoin.

—— (ed.) (1976) *Geography in Japan*, The Association of Japanese Geographers Special Publication No. 3, University of Tokyo Press.

—— Yamaga, S., Shimizu, K. and Inanaga, S. (eds.) (1964), *Nihon no toshika* (Urbanization of Japan), Tokyo.

Kobayashi, K. (1958) 'Quaternary Glaciation of the Japan Alps', *Jour. Fac. Liberal Arts and Science*, Shinshu University, *8*, 2. 13-67.

Komaki, S. (1937) *Senshi chirigaku kenkyu* (On the Essence of Prehistorical Geography), Naigaishuppan, Kyoto.

Kornhauser, D.H. (1979) *A Selected List of Writings on Japan Pertinent to Geography in Western Languages with Emphasis on the Work of Japan Specialists*, University of Hiroshima.

Kuwabara, T. (1976) *Chiseki-zu* (Cadastral Map), Gakuseisha.

Kyotodaigaku Bungakubu Chirigaku Kyoshitsu (Department of Geography, Faculty of Letters, University of Kyoto) (1982) *Chiri no shiso* (Thoughts in Geography), Chijinshokan.

Machida, T. (1960) 'Geomorphological Analysis of Terrace Plain — Fluvial Terraces Along the River Kuji and River Ara, Kanto District, Japan, *Sci. Rep. Tokyo Kyoiku Daigaku*, Sec. C., 64, 137-94.

Masai, Y. (1977) *Nichibei toshi no hikaku kenkyu* (Comparative Studies on Japanese and American Cities), Kokonshoin.

Matsui, I. (1931) 'Statistical Study of the Distribution of the Tonami Plain, Toyama Prefecture', *Japanese Journal of Geology and Geography, 9*.

Mino, Y. (1950) 'On the Gravel of Koromogawa, Akita Pref., Japan', *Sci. Rept., Tokyo Bunrika Univ.*, Sec. C., Vol 2, 91-106.

Miyakawa, Y. (1977) *Kogyo haichiron* (The Location of Industry in Japan), Taimeido.

Momiyama, M., (1947-50) 'Nihon ni okeru kakke no shippeichirigakuteki kenkyu' (Studies on Medical Geography of Beriberi in Japan), *Chirigaku Hyoron, 21*, 22-24.

—— (1977) *Seasonality in Human Mortality. A Medico-geograhical Study*, University of Tokyo Press.

Morikawa, H. (1974) *Chuchinchi kenkyu* (A Study on Central Places), Taimeido.

Nagaoka, A., Nakatoh, Y. and Yamaguchi, F. (eds.) (1979) *Nihon nogyo no chiiki kozo* (Regional Structure of Japanese Agriculture), Taimeido.

Nakano, T. (1954) 'Coastal Movement and Shoreline Development along the Pacific Coast of Japan since the Holocene Period – A Report of the Geomorphological Studies of Lowlands of Japan', *Bull. Geogr. Surv. Inst., 4*, 1, 87-113.

Nishimura, M. (1977) *Chushinchi to seiryokuken* (Central Places and Spheres of Influence), Taimeido.

Nishizawa, T., Yamashita, K. and Suzuki, H. (1979) 'Tokyo toshinbu ni okeru chion-bunpu' (Distribution of Ground Temperatures in the Central Area of Tokyo), *Chirigaku Hyoron, 52*.

Ohmori, H. (1978) 'Relief Structure of the Japanese Mountains and their Stages in Geomorphic Development', *Bull. Dept. Geogr. Univ. Tokyo, 10*, 31-85.

Oshima, G. (1972) *Suisanyoshokugyo no chirigakuteki kenkyu* (Geographical Study of Agriculture), Tokyo Daigaku Shuppankai.

Otsuka, Y. (1931) *Daiyonki* (The Japanese Quaternary), Iwanami-shoten.

Oya, M. (1977) 'Comparative Study of the Fluvial Plain Based on the Geomorphological Land Classification', *Geogr. Rev. Japan, 50*, 1-31.

Pinchemel, Ph. (1980) 'L'histoire de la géographie japonaise', *L'Espace géographique*, 165-71.

Sakaguchi, Y. (1961) 'Palaeogeographical Studies of Peat Bogs in Northern Japan', *Jour. Fac. Sci. Univ. Tokyo*, Sec. 2, 12, 412-513.
—— (1980) 'Characteristics of the Physical Nature of Japan with Special Reference to Land Form', in The Association of Japanese Geographers (ed.) *Geography of Japan*, Teikokushoin, 3-28.

Science Council of Japan (ed.) (1980) *Recent Trends of Geographical Studies in Japan*, Recent Progress of Natural Sciences in Japan, vol. 5.

Sekiguti, T. (1951) 'Studies on Local Climatology (VI) (VII)', *Pop. Met. Geogr., 2*, 170-88, 302-10.

Senda, M. (1974) *Umoreta minato* (Ancient Harbours), Gakuseisha.
—— (1980) 'Territorial Possession in Ancient Japan: The Real and the Perceived' in The Association of Japanese Geographers (ed.), *Geography of Japan*, Teikokushoin, 101-20.

Shiki, M. (ed.) (1975). *Nippon no hyoki no shomondai* (Problems on Ice Age in Japan), Kokonshoin.

Sugai, S. (1949) *Nihon no nogyo – Sono Keizaichirigakuteki Kenkyu* (Agriculture of Japan – An Economic Geographical Study), Kokonshoin.

Sugimura, A. (1950) 'Kantochiho-shuhen no Kaiteidankyu Sonota ni tsuite' (On the Submarine Terraces Along the Coast of Kanto Region and the Others), *Chirigaku Hyoron, 23*, 10-16.

—— (1967) 'Uniform Rates and Duration Period of Quaternary Earth Movements in Japan', *Journ. Geosc. Osaka City Univ., 10*, 23-35.

—— and Uyeda, S. (1973), *Islands Arcs*, Elsevier.

Suizu, I. (1969). *Shakaishudan no seikatsukukan* (Life-space of Society), Taimeido.

Suzuki, H. (1962) 'Klassification der Klimate von Japan in der Gegenwart und der letzten Eiszeit', *Japan Jour. Geol. Geogr., 33*, 221-34.

—— (1975) 'On the Utility of Monthly Data for the Dynamic Climatology of the World', *Bull. Dept. Geogr. Univ. Tokyo, 7*.

Suzuki, K. (1969) *Chiikikan kamotsu yusoro no keisoku to yosoku* (Measurement and Forecast of Interzonal Flow of Goods), Katsunihonsha.

Tada, F., Nakano, T. and Iseki, H. (1952) 'Shoreline Development of the Pacific Coast of Japan in Pre-historic Time', Proceedings of the 17th Congress IGU Washington, 386-91.

Takashige, S. (1975) *Kodai chusei no kochi to sonraku* (Arable Land and Villages in Ancient and Medieval Japan), Taimedo.

Takayama, S. (1974) *Kasenchikei* (Fluvial Geomorphology), Kyoritushuppan.

Takeuchi, A. (1973) *Nihon no kikai kogyo* (Geographical Study on Machinery Industry in Japan), Taimeido.

Takeuchi, K. (1974) 'The Origins of Human Geography in Japan', *Hitotsubashi Journal of Arts and Sciences, 15*, 1-13.

—— (1978) 'Some Remarks on the History of Regional Description and the Tradition of Regionalism in Modern Japan', *Hitotsubashi Journal of Social Studies, 10*, 36-44.

—— (1980) 'Geopolitics and Geography in Japan Reexamined', *Hitotsubashi Journal of Social Studies, 12*, 14-24.

Tanabe, K. (1971) *Toshi no chiiki kozo*, Taimeido.

Tanaka, K. (1957) *Shio to sakana no inyuro* (Transport Routes of Salt and Fish), Kokonshoin.

Tanioka, T. (1959) 'Le jori dans le Japon ancien', *Ann. Econ. Soc. Civil, 14*, 625-39.

—— and Yanaga, T. (eds.) (1979) *Isewan-gan chiiki no kodai jorisei* (The Jori-system in the Coastal Plain Around the Ise Bay), Tokyodo-shuppan.

Uchida, K. (1971) *Kinsei noson no jiukochiriteki kenkyu* (Population

Geography of Rural Settlement in the Late Middle Age), Chukokan.

Watanabe, H. (1968) *Jorisei no kenkyu* (A Study of the Jori System), Sogensha.

Watanabe, Y. (1955) *The Central Hierarchy in Fukushima Prefecture. A Study of Types of Rural Service Structure*, Science Report No. 4, Tohoku Univ. Geography.

Yada, T. (1975) *Sengo Nihon no sekitan sangyo – sono hokai to shigen no hoki* (The Coal Industry in Postwar Japan – Its Collapse and the Abandonment of Coal as a Natural Resource), Shinhyoron.

—— (1982) *Sangyohaichi to chiiki kozo* (Industrial Distribution and Regional Structure), Taimeido.

Yamada, Y. (1976) *Kodai Tohoku no frontia* (The North-eastern Frontier in Ancient Japan), Kokonshoin.

Yamaga, S. (1964) *Toshichirigaku* (Urban Geography), Taimeido.

Yamamoto, Sh. (1973) *Chagyo chiiki no kenkyu* (Studies on the Tea Cultivation and Production), Kokonshoin.

Yamamoto, S. (1953) *Chikasui chosaho* (Ground Water Survey), Kokonshoin.

—— (ed.) (1968) *Rikusui* (Inland Water), Kyoritsushuppan.

—— (1972) 'Current Studies on Hydrology in Japan', *Chirigaku Hyoron, 45*, 163-71.

Yamazaki, F. and Moritaki, K. (eds.) (1978) *Gendai Nihon no toshi supuraru mondai* (Problems of Urban Sprawl in Japanese Cities), 2 parts, Otsuki shoten.

Yamori, K. (1970) *Toshipuran no kenkyu* (The Study of City Plan), Taimeido.

—— (1974) *Toshizu no rekishi. Nihon hen* (History of City Maps of Japan), Kodansha.

Yatsu, E. (1955) 'On the Longitudinal Profile of the Graded Rivers', *Trans Amen Geographs Union, 36*, 655-63.

Yazawa, T. (1953) *Kiko keikan* (Climatic Landscape), Kokonshoin.

—— (1956) *Kikogaku* (Climatology), Chijinshokan.

Yazima, N. (1954) *Musashino no shuraku* (Rural Settlement in Musashino), Kokonshoin.

Yokota, T. (1967) *Tabako saibai chiikiron* (Geographical Study of Tobacco Growing Areas in Japan), Tokyo Keizai Shinposha.

Yonekura, J. (1949) *Shuraku no rekishichiri* (Geography of Settlement), Kokonshoin.

Yoshikawa, T. (1953) Continental Shelves Around the Japanese Islands, Actes du IV Congr. INQUA, Rome-Pise, Vol. 2, 554-62.

—— Kaizuka, S. and Ota, Y. (1964) 'Coastal Movements in the Late

Quaternary Revealed with Coastal Terraces on South-east Coast of Shikoku, South-western Japan', *Sokuchigaku Zasshi, 10*, 116-22.

—— Sugiura, A., Kaizuka, S., Ota, Y. and Sakaguchi, Y. (1973) *Nippon Chikei-ron* (Geomorphology of Japan), University of Tokyo Press.

Yoshimura, S. (1933) *Chiikikeisokuron* (Chorometrics), Iwanami-shoten.

Yoshimura, Y. (1942) *Chikasui* (Groundwater), Kawadeshobo.

Yoshino, M.M. (1957) 'Some Aspects on the Distribution of the Surface Winds Within a Small Area', *Kisho Shushin 75th Ann. Vol.*, 365-71.

—— (1961) *Shokiko* (Local Climate), Chijinshokan.

—— (ed.) (1971) *Water Balance of Monsoon Asia*, University of Tokyo Press.

—— (1975) *Climate in a Small Area*, University of Tokyo Press.

12 THE IBERIAN PENINSULA AND LATIN AMERICA

J. Vilà-Valentí

The principal purpose of this chapter is to review the evolution of geography first in Spain and secondly in the remainder of the Spanish-speaking world. Parallel trends in the Portuguese linguistic area are also reviewed, however, though more briefly.

Spain

Contemporary geography activity in Spain can be dated from the foundation of the Madrid Geographical Society (*Sociedad Geográfica de Madrid*) in 1876 (Vilà-Valentí, 1977). Since then, changes have been instituted both by natural scientists in the university Faculties of Sciences ('the natural science trend') and by teachers in secondary schools and Colleges of Education ('the pedagogic trend'). These largely occurred during the first third of the twentieth century. Since the Civil War (1936-9) developments have largely been promoted through the Faculties of Arts (*Facultades de Letras*) in the universities. Growth of the discipline since the 1940s reflects the work of the last group and from 1940 on few natural scientists (geologists and, to a lesser extent, biologists) played any significant part. Some support was obtained from various state organisations, however, notably the Geographical and Cadastral Institute (*Instituto Geográfico y Catastral*, now the *Instituto Geográfico Nacional*) and the Geological and Mining Institute of Spain (*Instituto Geológico y Minero de España*) which have provided cartographic and other sources of information.

The Development of Teaching and Research

Geography developed slowly in Spain until the middle of the 1960s, followed by a period of rapid expansion in both personnel and publications. Two clear stages can be identified, therefore, with characteristics indicated by the major chroniclers (Casas Torres, 1962; Capel, 1976).

The first stage, to 1965, was initiated in 1940 by the foundation of the Elcano Institute within the Council of Scientific Investigations (*Consejo de Investigaciones científicas*), the general Spanish organisation for stimulating scientific research. The Institute's journal, *Estudios Geográficos*, was launched in 1940 as the second Spanish geographical periodical – the first was the *Boletin de la Real Sociedad Geográfica*, published in Madrid since 1876.

Despite this development favouring the discipline, growth was very slow, reflecting the political and economic difficulties. By 1950 there were only five geography professors within the university Faculties of Arts. Desire for expansion was substantial, however, and an important introductory course was provided for undergraduates in 1946 (Vilà-Valentí, 1979-80). Three groups of geographers led this expansion: those in the University of Madrid and the Elcano Institute in Madrid, led by Amando Melón (1894-1975) and Manuel de Terán (1905-); in the University of Zaragoza and the local section of the Elcano Institute, under José Manuel Casas Torres (1918-); and a smaller group in the Barcelona section of the Elcano Institute, led by Lluis Solé Sabarís (1908-). To the chairs of geography already existing at Madrid, Barcelona, Valencia, Valladolid and Zaragoza others were added, at Santiago de Compostela, Oviedo, Granada, Murcia and Salamanca, with second professors at Madrid and Zaragoza.

The second stage, from 1965 on, was characterised by rapid expansion in the numbers of university teachers and students of geography. By 1982 there were qualified staff teaching geography students at all twelve traditional Spanish universities, at the new universities (Alicante, Autonomous of Barcelona, Autonomous of Madrid, Córdoba, Extremadura, Málaga, Palma de Mallorca and Santander) and at certain other colleges (such as Tarragona, affiliated to the University of Barcelona). In total, there were some 25 geography groups in either the Faculty of Arts (the traditional affiliation), in a Geography-with-History Faculty (the more recent organisational mode), or in affiliated colleges.

Research is in progress among all these groups, and several have established new journals. *Geographica* has been published since 1954 by geographers at the University of Zaragoza; *Cuadernos de Geografía* was launched in Valencia in 1964; and in 1967 the University of Barcelona commenced production of *Revista de Geografía*. Others have followed and by the 1980s some 15 academic journals were being published.

Teaching was initially in joint degree programmes. The first full degree course in geography was only established in 1969, at the University of Barcelona; the first students graduated in 1972. Other universities introduced special degrees or other courses shortly afterwards.

The Preponderant French Influence

The two stages identified above are distinguished in a variety of ways. For example, the first stage was characterized by the dominant French influence, with regard to methods of investigation. During the 1920s the Spanish geographer-teacher Pau Vila (1881-1980) became aware of the novel, scientific features of French geographical studies of the time, in contrast to the descriptive work of previous decades. In the prologue to his regional study of the border region of La Cerdanya (published in Catalan in 1926) he wrote:

> In this volume we are starting a new type of study . . . La Cerdanya is the first stone in the study of regional geography . . . We have been guided, above all, by the modern geographical currents of the French school, directed by the late master P. Vidal de la Blache . . . We can confidently say that the geographical concept which our collection advances will be the interpretation of geography as a natural science . . . Geography is a science of scientific maturity, when considered that, more than many other sciences, it uses materials of related sciences together with its own material for its synthesis (pp. 9-11).

This French mode of 'doing geography' was seen as clearly conducive to a logical order and coherence of thought, in contrast to earlier work which comprised descriptive, often disjointed, accounts of accumulated information.

The preponderant influence of the French school was in part occasioned by direct contacts. The most notable example of this is Pierre Deffontaines (1894-1978), who spent 25 years in Barcelona as director of the French Institute. Subsequently, some Spanish geographers studied in France.[1] However, the main source of ideas came through the geographical journals and by the translation of important French works into Spanish. The second (abridged) edition of Jean Brunhes's text on human geography was published in Castilian in 1948, for example, as later were works by A. Demangeon, D. Faucher and E. de Martonne. Among post-war French authors, the works of Pierre George in agricultural, economic and urban geography have been made available in translation.[2]

The dominance of the French school did not go unchallenged. Some geographers — such as Emili Huguet del Villar (1871-1955), José Gavira (1903-51) and L. García Sáinz (1894-1965) — were influenced by

German work. However, this had little effect on Spanish geography as a whole. One German geographer – Hermann Lautensach – visited and worked in the Iberian Peninsula from 1927 on, for example, and his major work was translated into Castilian in 1964, but he had no noticeable influence on the thought and methodology of Spanish geographers.

The preponderant French influence implies a preference for regional analysis, concerned with the concept of 'the geographical region' as developed by Vidal de la Blache and interpreted in the doctoral theses of his students. It was commonly believed that the training of an academic geographer culminated with the preparation of a doctoral thesis in regional geography. This involved detailed analyses of the chosen area followed by a synthesis (as indicated in the quote above from Pau Vila's work). On the value of synthesis in regional geography, the editor of *Geographica* wrote in the first issue of the journal:

> Synthesis, the overall view, reconstructing the picture and establishing the connection and mutual interdependencies between the phenomena studied in isolation by each of the individual sciences, encompassing with a single global view a living reality ... is ... the magnitude and the risk of the geographer's task. (Casas Torres, 1954, p. 3.)

The first Spanish thesis set in this mould was presented by Salvador Llobet at the University of Madrid in 1944. He conducted a similar analysis in Andorra, and both were published in 1947. During the next 20 years most doctoral theses followed the French model, and included Alfredo Floristán on las Bárdenas (Navarre, 1949), Manuel Ferrer on the countryside of Cariñena (Aragón, 1954), J. García Fernández on la Alcarria (Castile, 1954), J. Vilà Valentí on el Bages (Catalonia, 1956), Vicente Rosselló on south-east Mallorca (1962) and Maria de Bolos on Olot (Catalonia, 1966). (For a full list referring to the University of Madrid see 'Tesis doctorales', 1981.) Most of these dealt with *comarcas*, areas comparable in extent and complexity to the French *pays* rather than the region.

Thematic Diversification

The dominance of the French regional model began to wane in the 1960s. Non-regional themes had been investigated earlier, especially in human geography (as illustrated by Joaquín Bosque's thesis on the urban geography of Granada in 1955), but it was only in the 1970s that such diversification was widely reflected at the highest levels – i.e.

in degree theses, especially those for doctoral degrees. This marked the beginning of the second stage.

At the start of the 1960s Spanish geography was still clearly dominated by French influence, transmitted by those trained immediately after the Second World War. At that time thematic specialisation was increasingly common in France (as in the works of Birot, George and Tricart), and such a tendency developed in Spain too. Thus there was work in geomorphology, led by Lluís Solé Sabarís; in population geography (J.M. Casas Torres); agricultural geography (with several leaders); and urban geography (Manuel de Terán); associated with them were Salvador Mensua (geomorphology); Manuel Ferrer (population geography); Angel Cabo, J. García Fernández and J. Vilà-Valentí (agricultural geography); and Joaquín Bosque (urban geography).

This thematic trend strengthened during the 1960s. Other specialities were added. Climatology was developed in the pioneer works of A. López Gómez, and in the translation into Castilian of two major texts (that by Köppen in 1948 and that by Austin Miller in 1953); it was later advanced at the University of Barcelona, led by Luis Miguel Albentosa, and at Granada, led by J.J. Capel Molina. Similarly, the study of landscape was advanced through careful geomorphological, biogeographical and agricultural analyses. The influence of Georges Bertrand (University of Toulouse le Mirail) was important in this, his terminology and approach being adopted by groups led by Maria de Bolòs (University of Madrid) and E. Martínez de Pisón (University of La Laguna and University of Madrid). Physical geography was also developed in Valencia (under V. Rosselló), population geography at Palma de Mallorca (B. Barceló), and urban geography and regional planning at Granada (J. Bosque).

The concept of the functional region was also introduced, through the influence of several Anglo-Saxon geographers and economists (Vilà-Valentí, 1971, 1973) and that of French workers (P. Claval; E. Juillard; J. Labasse). Not only did this involve a break from the traditional concept of the formal region, but it moved towards a more applied orientation, with emphasis on topics such as organisation of territory and administrative division of territory. Two theoretical works focused on this development of spatial and regional analysis (Solé Sabarís, 1975; Vilà-Valentí, 1980).

These changes are readily appreciated by investigation of the journal literature in, for example, *Estudios Geográficos* (Madrid), *Revista de Geografía* (Barcelona), *Cuadernos Geográficos* (Granada) and the second series of *Geographica* (Madrid) (Río, 1975). Similar themes are indicated in the collected papers presented at International Geographical Congresses and in *festchrift* volumes (e.g. *Aportación española*, 1964, 1968; *Homenaje Melón*, 1966; J.M. Casas Torres, 1972; 'Homenaje Terán', 1975; *Boletín de la Real Sociedad Geográfica*, 1976; 'Homenaje Solé', 1979-80). In addition, several foreign texts (in geomorphology: Dernau; climatology: Viers and Durand-Dastès; biogeography: Lacoste and Salanon; agricultural geography: Clout, Morgan and Munton; urban geography: Johnson; and spatial organisation: Labasse) were translated, influencing the whole of the Spanish-speaking world.

Contemporary with this increased work on thematic topics, some geographers developed an interest in the history and philosophy of geography.[3] The critical evaluation involved was clearly linked to the renovation of geographical scholarship taking place, and was closely associated with the new influences and approaches.

The 'New Geography' and the Anglo-Saxon Influence

Clear signs of a further reappraisal, with the introduction of both theoretical geography and quantitative methods, can be identified in the late 1960s and early 1970s. These new ideas were introduced via reading the relevant literature (especially the publications of Bunge, Chisholm, Chorley, Haggett and Schaefer) and via personal contacts, many of which were forged at International Geographical Congresses (notably London in 1964) and through courses in and short visits to Great Britain and the USA.[4] It was mainly the young graduates and lecturers who responded to this 'new geography' with enthusiasm, setting up a generational factor which can lead later to some disenchantment.

The new work initially was centred at only a few universities. At the University of Barcelona, for example, lectures and comments on theoretical geography began in 1968/9, with several seminars conducted in accord with the new methods (Capel, Tatjer and Batlori, 1970); Schaefer's article was published in translation in 1971, and the first part of a work on the history and characteristics of theoretical geography was prepared (Vilà-Valentí, 1971). Meanwhile, E. Lluch at the Autonomous University of Barcelona compiled a translation of Chisholm's works into Castilian in 1969. And in 1971 the first issue of the second series of *Geographica* (published by the Institute of Applied Geography at Madrid) presented and discussed the new ideas.

This trend represents a profound change in the nature of overseas influence on Spanish geography. In particular, the Anglo-Saxon countries replaced France as the major source of new ideas.[5] But the French influence did not entirely disappear.[6] Nor was Spanish work merely a copy of foreign ideas. Instead, efforts were made by various leaders (Casas Torres, Manuel de Terán) to adapt approaches developed in the first stage, and these were continued by their followers (J. Bosque, A. Cabo, J. García, Fernández, M. Ferrer, A. Floristan, A. López Gómez, S. Mensua, J. Vilà-Valentí). And regional work continued, albeit with innovations in the concept of the functional region and the techniques of analysis.[7]

The boom in theoretical geography is indicated by the number of Anglo-Saxon authors whose work was translated: they included Berry, Blaut, Chisholm, Chorley, Clout, Gould, Ginsburg, Haggett, Harvey, Morgan, Peet, Schaefer and Zelinsky (see also Vilà-Valentí 1973, 1983). By 1976 the ideas were spread through the Spanish universities, as indicated, for instance, by a sort of 'manifesto' from Palma di Mallorca (Quintana, 1976). Doctoral theses in the 'new geography' were produced (Vilà-Valentí, 1983) and a text on quantitative methods in Castilian was published by a Spanish and an English author working in collaboration (Estébanez and Bradshaw, 1979).

Pluralism in the 1970s

With these developments it is clear that Spanish geography had lost the uniformity that characterised it in the 1950s. It is now characterised by a pluralism that incorporates (Vilà-Valentí, 1981):

(1)   A continuation of the traditional regional themes, in some cases with objectives and methods slightly modified in the context of the 'new geography'.

(2)   A preference for integration as a consequence of new methods (as in the concept of the integrated landscape: see the conference on 'Geosystem and landscape' held at Barcelona in 1980 and reported in *Collección geosistema*, 1983); at the same time, for some authors the influence of the 'new geography' was restricted to the introduction of quantitative methods.

(3)   Work clearly located within the Anglo-Saxon concept of theoretical geography, including quantitative analysis.

(4)   A diversification of content, approaches and aims, in part as a reaction to the 'new geography', its instrumentalism and lack of human content. Themes studied include environmental depreciation and perception of and reactions to environments. Human behaviour is being analysed and relevant contacts developed with other disciplines (such as ecology and social psychology). Applied

studies are flourishing, as are those of the philosophy of geography (as in the works of H. Capel, M.D. García Ramón, J. Vilà-Valentí; see the work of the Madrid group in Gómez Mendoza, Muñoz and Ortega, 1982).

(5) Open criticism and attempts to achieve a profound transformation of geography's subject matter. Such studies have clear political implications, are related to Anglo-Saxon radical geography and criticise the false 'neutrality' and 'ineffectiveness' of quantitative work. Good examples are provided in the Barcelona journal *Geocritica* (published since 1976). The work is generally grounded in Marxist ideals (Ortega, 1977; García Ramón, 1978), although there are other influences, such as the libertarian ideology based on the work of Reclus. The critical attitude has developed most strongly among students, who have demanded a thorough revision of the concepts and teachings of geography (at conferences held annually, since 1978, at Barcelona, Granada, Salamanca, Madrid and Cáceres: C. García, Les and Roca, 1980).

Some of these changes are new and it is difficult to identify their real significance for the future. Nevertheless it is clear that in Spain geography has developed in ways and to an extent that was not conceivable even three decades ago. As an institutionalised discipline, it is severely restricted (in its competition with other disciplines for a place in the curricula and for teaching and research posts). Internally, however, it is both lively and varied.

Portugal

To some extent, contemporary developments in Portuguese geography are, not surprisingly, somewhat similar to those in neighbouring Spain. The two countries have similar historical and cultural traditions, reflecting commercial and colonial influences, and these were reflected in Portugal in the foundation of the Geography Society of Lisbon (1875) and the improvements in state cartographic and statistical services. At the beginning of the twentieth century, developments in the natural sciences and in education, similar to those in Spain, fostered the development of geography and the creation of some university lectureships. In the 1920s and 1930s, however, Portuguese geography developed unique characteristics, as consequences of the influence of two important personalities.

The first of these two geographers – A. de Amorim Giráo (1895-1960) – was (with the exception of F.X. da Silva Telles, 1860-1930) the only Portuguese scholar to adopt completely the French regional approach before the Second World War. Professor of Geography at Coimbra, Giráo published a regional study of the Vouge basin in 1922, and one of the town of Viseu in 1925.

Although the French influence was strong, via Giráo, a unique quality was added to Portuguese geography through the impact of the German Lautensach. His work on the Iberian Peninsula, mentioned above, was especially focused on Portuguese territory and he made a deep impression on the country's small group of geographers. Prominent among these from the end of the Second World War was Orlando Ribeiro (1911-). His basic approach stemmed from the work of de Martonne and Demangeon, but he assimilated Lautensach's approach too in the *Centro de Estudios Geográficos*, which he directed at the University of Lisbon for nearly four decades. Within this, the French influence increased during the 1950s through the presence of Pierre Birot.

Developments in Portuguese geography at Lisbon and at Coimbra (the latter directed by Alfredo Fernandes Martins, 1916-) are illustrated by the 16th International Geographical Congress held at Lisbon in 1949. Orlando Ribeiro was the general secretary to the Congress, and A. Fernandes Martins and Mariano Feio played major roles in the activities and publications. The *Centro de Estudios Geográficos* became more prominent after this event. From 1960 on it has published a series of monographs under the title *Chorographia*. It took a major interest in the study of Portuguese African territories, as examples of tropical environments, and several of its scientific projects, including doctoral theses, were supported by the *Junta de Investigacoes Científicas de Ultrumar*.[8]

Renovation Since the 1960s

The focus on overseas territories continued until Portugal lost its African possessions in 1975, giving a particular individuality to the study of geography there. Expansion was linked to groups of geographers at the universities in Luanda and Lourenço Marques. But in Portugal itself growth was limited, with groups of geographers at Lisbon, at Coimbra (now led by J.M. Pereira de Oliveira) and, from 1972 on and linked to the Lisbon group, at Oporto, and later at Evora.

Most of what has happened in Portuguese geography in recent decades is linked to developments at Lisbon's *Centro de Estudios*

Geográficos, which has published the journal *Finisterra* since 1966. Two main trends stand out. First, there has been thematic diversification, accompanied by a greater depth of study in certain specialisations. This has occurred in several branches of physical geography, especially geomorphology (the theses of Antonio de Brun Ferreira and M. Eugénia S. Moreira Lopes), and climatology (led by Suzanne Daveau, a French teacher linked to the University of Lisbon in the 1960s: 1981). In human geography, the emphasis has been on urban and agricultural studies; numerous new methods and approaches have been applied to the former in Lisbon (Gaspar and Gama, 1981; Gaspar and Ferráo, 1981) whereas more traditional methods have been used in the latter at Coimbra. In regional geography, in which the thesis by Carminda M. Cavaco in 1976 stands out, the trend has been similar to that in Spain of adopting developments in the French school.

The second major trend in Portuguese geography has been the application of theoretical and quantitative methods. Led by Jorge B. Gaspar, this has provoked a profound change in concepts and methodology. Gaspar's career was greatly influenced by his studies and contacts in Sweden, so that the introduction of these ideas followed a separate route from the contemporaneous development in Spain. Gaspar's thesis on the city of Evora was published in 1972.

By the end of the 1970s, therefore, the Lisbon group, under the guidance of O. Ribeiro, S. Daveau, I. do Amaral and J. Gaspar and assisted by C.A. Medeiros and C. Cavaco, had considerable strength and diversity of geographical concepts and methods available. Unlike the situation in Spain, however, there has been no development of a critical attitude, even though political and cultural changes may have favoured it. It may be that this is because study and teaching of geography takes place on a much smaller scale in Portugal than in Spain.

There has been increasing contact between Spanish and Portuguese geographers in recent years. Previously, the main contacts were at a personal level and involved O. Ribeiro. There was a joint meeting at Jaca in 1946, however (Vilà-Valentí, 1979-80), and there is now a series of fruitful exchanges at the Coloquios Ibéricos de Geografía (*I Coloquio Ibérico de Geografía*, 1981; Medeiros, 1980).

Latin America

Generalisations about developments in Latin America are difficult to achieve. There is a great diversity apparent, depending on both individuals and contrasting social and economic contexts. Further, the nature of the outside influences varies somewhat, although in many of the countries there have been frequent, direct contacts with geographers in the United States of America.

In the 1920s Latin American geography was dominated by a traditional approach and descriptive exposition of regional study, alongside some more modern approaches. The latter were introduced mainly by Americans and Germans, some of whom worked in Latin America for several years: among the Germans were Carl Troll, who worked in Colombia, Walther Penck, who worked in Argentina, and Leo Waibel, who worked in Brazil and Central America. Most of these outside workers studied either general features of the tropical world, such as climate and vegetation, or the historical and socio-economic characteristics unique to Latin American countries.

The Formation of National Groups

After the Second World War several Latin American countries developed stable groups of local geographers pursuing particular interests. The creation of these groups was stimulated in several cases by the presence of foreign researchers and teachers, and in others by personal links developed in other countries (often during periods of study abroad) by Latin Americans. Thus, for example, the Germans Wilhelm Rohmeder and Fritz Kunz were influential in Argentina; the Frenchman Jean Borde, from Bordeaux, worked and taught in Chile, while Pierre George, Jean Tricart and Olivier Dollfus travelled and maintained direct contacts in Argentina, Brazil, Chile, Peru and Venezuela; the Catalan, Pau Vila, had considerable influence in Venezuela; and another Spanish exile, Angel Rubio, started a small group of geographers in Panama. (For more detail on foreign influence, see Reboratti, 1982.)

These varied outside influences, together with internal educational and cultural improvements, led to the establishment of national schools of geography. Many of these were led by one notable personality, such as Jorge Vivó (1905-80) in Mexico and Humberto Fuenzalida (1904-66) in Chile (Santis and Gangas, 1982). Other leaders were somewhat less influential, such as Salvador Massip in Cuba, Rafael Picó in Puerto Rico, Jorge Chebataroff and Ignacio Martínez Rodriguez in Uruguay, and Federico Daus and Mariano Zamorano in Argentina.

Four Hispano-American countries stand out during this period of geographical growth. (The case of Brazil is considered below.) In Mexico, development focused on the College and Institute of Geography at the National Autonomous University; Venezuela had centres at Caracas (Central University of Venezuela and the Education Institute) and Mérida (University of the Andes); in Argentina there were groups in Buenos Aires, Mendoza (University of Cuyo), La Plata and Tucuman; and in Chile the main centres were the University of Chile and the Catholic University. In all of these, the main emphases were descriptive, usually local, regional studies, using approaches to thematic analyses reflecting a variety of origins but concentrating on geomorphology, agricultural and urban geography. There was little attention to geographical thought. Instead, many studies had a strong applied bias, inspired by the need for economic growth, socio-economic development and better use of natural resources (Vilà-Valentí, 1968) and aided by technical advances in the USA in cartography, aerial photography, photogrammetry, prospective analysis and statistics. Many Latin American geographers have been sensitive to such problems because of the influence of sociologists and economists from widely different ideologies, such as Raul Prebisch and Gunder Frank.

Interest in geography was increased through the educational systems and a concern for teaching the subject at various levels. This is illustrated by the production, through the Commission on Education of the International Geographical Union, of a series of essays involving the collaboration of some 30 Latin American geographers (Brouillette and Vilà-Valentí, eds., 1975). Several of the main national geographical associations have been concerned with education as have some of the main national journals: *Anuario de Geografía* (Mexico since 1961); *Revista Geográfica* (Mérida, Venezuela, since 1959); *Informaciones Geográficas* (Santiago de Chile since 1951); *Gaea* (published by the Argentinian Society of Geographical Studies, Buenos Aires) and *Boletín de Estudios Geográficos* (Mendoza, Argentina, since 1948). The Pan-American Institute of Geography and History has stimulated much contact among Latin American geographers, notably via its two publications *Revista geográfica del Instituto Panamericano de Geografía e Historia* (IPGH) (since 1941) and the *Boletín Aéreo del IPGH* (since 1955).

Recent Developments

The different national groups have evolved in a variety of ways during the last 15 years. Several have consolidated and expanded, but some have experienced setbacks and crises, partly, no doubt, as a consequence of problematical political and economic contexts (Reboratti, 1982). Two general developments can be identified, however. On the one hand there has been diversification of topical interests, in many cases focused on applied themes such as those related to ecological problems, socio-economic development, and territorial organisation or regional planning. On the other hand, there have been considerable improvements in technical knowledge involving only a few geographers in some countries. The IPGH has contributed to the latter (relating to skills in photogrammetry and quantitative and cartographic methods) by creating a Pan-American Centre of Geographical Studies and Investigations (CEPEIGE) in Ecuador in 1973; in July and August of that year, the 'Primer Curso Internacional de Geografía Aplicada' was held, emphasising the teaching of methodology.

Despite these advances, many geographers exhibit either no appreciation or a faulty one of the developments, and their work has been confined to descriptive studies of little value or relevance.[9] Some Latin American geographers, notably those who have studied abroad (especially in the USA), are aware of this problem. Only in a few of the countries have the concepts of theoretical geography been introduced, however, and then only recently: the first issue of *Síntesis Geográfica* (Caracas, 1977) indicated that 'special emphasis would be given to those articles which used the new methodology and showed tendencies towards the New Geography'. Even so, the new concepts, having had no time to mature or adapt, have already been subjected to sharp criticism by minority groups. Some radical attitudes have been introduced, reflecting not only generational factors but also the realities of Latin America which lead to criticisms of attitudes considered neutral and demands for geographical analyses that can 'effect a profound change and promote socio-economic development' (G. García 1975, p. 81).

Many geographers are relatively unaffected by these changes and critiques, however. They have been content to accept a few conceptual and methodological changes within the context of theoretical geography. As a result, their published works clearly favour traditional topics and approaches, with certain thematic and methodological innovations and fairly remote pragmatic or applied objectives.

Brazil

Brazil deserves separate attention because of its isolation from the Spanish-speaking area and the considerable developments in geography there.

A fairly large group of Brazilian geographers was established in the 1950s, under the influence of French teachers (including P. Deffontaines, P. Monbeig and F. Ruellan) plus the German Leo Waibel. Several departments were established in the universities (Rio de Janeiro, São Paulo, etc.), alongside the important research centre, the *Instituto Brasileiro de Geografia e Estatistica*, founded in Rio in 1938 (Bernardes, 1982). Leaders among the geographers included F. de M. Soares Guimâres (1906-79) and Aroldo de Azevedo (1910-74). Brazil thus had the largest, most diversified and probably most dynamic group of geographers in Latin America at the time, with a flourishing and large *Associacão dos Geógrafos Brasileiros*, several journals (including *Revista Brasileira de Geografia* – Rio, since 1939 – and *Boletin Paulista de Geografia* – São Paulo, since 1949) and some monograph series (such as the *Noticia geomorfológica*, first published in 1958).

The post-war generation of Brazilian geographers (including H. O'Reilly Sternberg, Nilo Bernardes, Lysia M. Cavalcanti and Orlando Valverde) had an open and receptive attitude to the discipline, and shared a concern for the education of geography teachers and researchers. Their works focused on analyses of regional landscapes and on agricultural and urban topics.[10]

From the 1970s on, developments in theoretical geography began to take root and some quantitative methods, including those well outside traditional frameworks, diffused rapidly. Under the direction of Antonio Christofoletti, a group at the University of Rio Claro has formed a centre of progressive study, formalised as the *Associacão de Geografia Teorética*: a *Boletín de Geografia teorética* has been published since 1971 (see the important review by Christofoletti, 1976).

The regional Latin American Geographical Conference (organised by the International Geographical Union (IGU)) held in Rio in August 1982 illustrated the strengths and diversity of Brazilian geography. Its theme was one which has preoccupied Latin American geographers for many years – 'the analysis of the problems of the Third World, in particular, their relationship with geographical space'.

Notes

1. Montserrat Rubió studied in Grenoble in 1949 with Raoul Blanchard and Paul Veyret; Alfredo Floristán and J. Vilà-Valentí studied in Bordeaux with Louis Papy and Henry Enjalbert (in 1950 and 1954 respectively). Floristán (1955) later published a significant study of the French school.
2. The introductory texts by André Allix (1950) and by Pierre Gourou and Louis Papy (1967) were translated and widely used. Thematic texts available in translation included those on physical geography by Pierre Birot (1962), on human geography by Max Derruau (1964), and on urban geography by Georges Chabot and Jacqueline Beaujeu-Garnier (1970).

3. Only a few papers on this topic appeared in the 1940s and 1950s (notably by Manuel de Terán and J.M. Casas Torres), although various papers presented in the entrance examinations for university teaching posts are of interest. The history of geography, in the traditional sense, was substantially developed by Amando Melón (a large number of his papers are reprinted in *Homenaje Melón*, 1966).

4. Among those with personal contacts in the UK and USA were J. Bosque, J. Estébanez, M.D. García Ramón, E. Lluch and J. Vilà-Valentí.

5. Horacio Capel, professor at the University of Barcelona, illustrates this in his introduction to the translation of Schaefer's paper:

Our attention will be directed towards the achievements of schools of geography until now little known in Spain (British, North American, Scandinavian, German, Russian and Polish), thus counteracting the unilateral influence the French school has traditionally had in our country (Capel, 1971, 13).

6. The relationship continued in geomorphology (with a preference for the ideas and methods of P. Birot and J. Tricart), climatology (P. Pédelaborde; C.P. Peguy), landscape study (C. Bertrand), historical development (F. Braudel), social and economic geography (P. George), agricultural geography (R. Brunet; J. Bonnamour), urban geography (J. Beaujeu-Garnier) and the theory of geography and the history of geographic thought (P. Claval).

7. A large number of regional theses were completed after the 1960s – for example, at the universities of Granada, Salamanca, Santiago de Compostela and Valencia, where the tradition was maintained that Spanish geographical research should primarily be concerned with regional and sub-regional (*comarcas*) studies (Terán, 1960).

8. Examples incude the doctoral theses of Raquel Soeiro de Brito on the Island of S. Miguel (1955), Francisco Tenreiro on S. Tomé Island (1961), and Ilidio do Amaral on the Island of Santiago de Cabo Verde (1963); later works considered Angola and Mozambique (e.g. Amaral, 1981, and the works of G.A. Medeiros).

9. One Latin American geographer has written that

On making an analysis of the articles, magazines and geography books which are published in Latin America the continuing predominance of descriptive studies of single phenomena can be noted, with little possibility of application in a general development plan of a particular area, region or country . . . Geography, in many of our countries, still suffers from the illness of description . . . , of superficial description (Garcia, G., 1975, pp. 79-83).

10. The Brazilian geographers named organised the International Geographical Congress at Rio in 1956, which produced contacts with geographers from many parts of the world who had direct interests in tropical areas (see Monteiro, 1980). The 1982 Regional Conference was organised by Speridiao Faissol.

References

Only the works which refer directly to the themes discussed are indicated. Of the remaining geographical publications, we only mention some collective works which are very important and can easily be consulted, in order to restrict the length of the references.

Amaral, I. do (1981) 'Notas acerca do ensino e da *investigaçáo cientifica em Geografia, en Portugal*' ın *J. Coloquio Ibérico de Geografía*, Universidad de Salamanca, 23-33.
Aportación espanola al XX Congreso Geográfico Internacional (1964) Instituto Elcano de Geografía and Instituto de Estudios Pirenaicos, Madrid-Zaragoza-Barcelona.
Aportación española al XXI Congreso Geográfico Internacional (1968) Instituto de Geografia Aplicada, Madrid.
Bernardes, N. (1982) *Alocuçao do encerramento da II Conferencia Regional Latinoamericana*, Rio de Janeiro.
Boletín de la Real Sociedad Geográfica (1976) (*112*) Spanish papers to 23rd International Geographical Congress, Moscow, 2 vols., Madrid.
Brouillette, B. and Vilá-Valentí, J. (eds.) (1975) *Geografia de América latina. Métodos y temas monográficos*, Teide, Barcelona, on behalf of UNESCO.
Capel, H., Tatjer, M. and Tatllori, R. (1970) 'La población básica de las ciudades españolas', *Estudios Geografícos, 31*, 29-79.
—— (1971) 'Schaefer y la nueva Geografía' in Schaefer, Spanish translation *Exceptionalism in Geography*, 5-13.
—— (1973) 'Percepción del medio y comportamiento geográfico', *Revista de Geografía, 7*, 58-150.
—— (1976) 'La Geografía'española tras la Guerra Civil', *Geocritica*, 1.
—— (1982) *'Filosofia y Ciencia en la Geografía, contemporánea'*, Barcanova, Barcelona.
Casas Torres, J.M. (1954) 'La Geografía aplicada', *Geographica, 1*, 1, 3-9.
—— (1962) 'Lo stato attuale degli studi geografici in Spagna', Società Geografica Italiana, Roma. Spanish translation with some modifications: 'Estado actual de los estudios geográficos en España', in Aportación española al XX Congreso Geográfico Internacional (1964), Instituto Elcano de Geografía and Instituto de Estudias Pirenaicos, Madrid-Zaragoza-Barcelona, 275-87.
—— (1964) *Las fronteras de la nueva Geografía*, Universidad de Zaragoza, Zaragoza.
Christofoletti, A. (1976) 'As características da Nova Geografia', *Geografía*, 1, 3-32.
Coloquio sobre geosistema y paisaje (1983) Departamento de Geografia Universidad de Barcelona.
Daveau, S. (1981) 'Investigações de climatologia no Centro de Estudos

Geográficos de Lisboa' in I *Coloquio Ibérico de Geografía*, Universidad de Salamanca, 47-54.

Estébanez, J. and Bradshaw, R.P. (1979) *Técnicas de cuantificación en Geografía*, Tebar Flores, Madrid.

Floristán, A. (1955) 'Los estudios geográficos en Francia', *Geographica, 2*, 9-20.

García, C., Les, R.J. and Roca, J. (1980) 'Los dos primeros Encuentros de estudiantes de Geografía (1978, 1979). Una reflexión entorno de los problemas de la Geografía española', *Revista de Geografía, 14*, 107-14.

García, G. (1975) 'Otra Geografía latinoamericana. Algunas reflexiones críticas en torno a la metodología', *Revista de Geografía, 9*, 79-90.

García Ramón, M.D. (1978) 'La Geografía radical anglosajona', *Documents Anàlisi Metodològic Geografia, Universitat Autònoma Barcelona, 59-69*.

Gaspar, J. and Ferrâo, J. (1981) 'As cidades portuguesas e a Geografia Urbana na Universidade de Lisboa' in *I Coloquio Ibérico de Geografía*, Universidad de Salamanca, 189-98.

—— and Gama, A. (1981) 'Perspectivas de Geografía Humana en Portugal: Ensino, investigaçâo e carreiras' in *I Coloquio Ibérico de Geografía*, Universidad de Salamanca, 67-78.

Gómez Mendoza, J., Munoz, J., and Ortega, N. (1982) *El pensamiento geográfico*, Alianza Editorial, Madrid.

'Homenaje a D. Manuel de Terán' (1975) *Estudios Geográficos, 36.*

Homenaje al Excmo. Sr. Amando Melón y Ruiz de Gordejuela (1966) Instituto de Estudios Pirenaicos and Instituto Elcano de Geografía, Zaragoza.

'Homenaje a Luís Solé Sabarís'(1979-1980), *Geographica, 21-2.*

I Coloquio Ibérico de Geografía,(1981), Universidad de Salamanca, Salamanca.

José Manuel Casas Torres. Homenaje a una labor (1972) Instituto de Geografía Aplicada, Madrid and Zaragoza.

Medeiros, C.A. (1980) 'Os dois primeiros Colóquios Ibéricos de Geografía', *Finisterra, 15*, 261-6.

Monteiro, C.A. de F. (1980) *A Geografia no Brasil (1934-1977), Avaliaçâo e tendências*, Universidade de Sâo Paulo, Sâo Paulo.

Ortega, N. (ed.) (1977) *Geografías, Ideologías, Estrategias espaciales*, Dédalo Ediciones, Madrid.

Quintana, A. (1976) 'Por una crítica teórica de la Geografía', *Mayurqa*, 209-23.

Reboratti, C.E. (1982) 'Human geography in Latin America', *Progress* in *Human Geography*, 397-407.

Ribeiro, O. (1972) 'Nueva Geografía y Geografía clásica. A propósito de dos publicaciones recientes', *Revista de Geografía, 6*, 145-67.

Río, Ma del (1975) 'La geografía en Espana desde 1940 a 1972, a través de las principales revistas geográficas' in 'Homenaje a D. Manuel de Téran', *Estudios Geográficos, 36*, 1033-48.

Santis, H. and Gangas, M. (1982) 'Notas para la historia de la *Geografía,* contemporánea en Chile (1950-80)', *Revista de Geografía, 16*, 5-21.

Schaefer, F.K. (1971) *Excepcionalismo en Geografía,* (translation and commentaries of H. Capel), Universidad de Barcelona, Barcelona.

Solé Sabarís', Ll. (1975) Sobre el concepte de regió geogràfica i la seva evolució', *Miscel·lània Pau Vila*, Societat Catalana Geografia, Montblanc-Martin, Granollers, 413-74.

Terán, M. de (1957) 'La causalidad en Geografía Humana. Determinismo, posibilismo, probabilismo', *Estudios Geográficos, 18*, 273-308.

—— (1960) 'La situación actual de la Geografía y las posibilidades de su futuro', *Enciclopedia Labor, 4*, 23-39.

'Tesis doctorales de Geografía defendidas en la Facultad de Filosofía y Letras de la Universidad Complutense de Madrid desde el curso 1934-35' (1981) *Anales de Geografía de la Universidad Complutense*, I. Sección Geografía, 347-55.

Venturini, V., O.L. (1977) 'Epistemología de la Geografía', *Síntesis Geográfica, 1*, 3-8.

Vila, M.A. (1981) 'La tasca geogràfica de Pau Vila a Amèrica', *Revista de Geografía, 15*, 17-22.

Vilà-Valentí, J. (1968) 'Algunos puntos de vista acerca de la Geografía aplicada', *Revista de Geografía, 2*, 43-55.

—— (1971 and 1973) '¿Una nueva Geografía?', *Revista de Geografía, 5*, 5-38; 7, 5-57.

—— (1977) 'Origen y significado de la fundación de la Sociedad Geográfica de Madrid', *Revista de Geografía, 11*, 5-36.

—— (1979-80) 'El curso de Geografía general y del Pirineo (Jaca, 1946)' in 'Homenaje a Luis Solé Sabarís', *Geographica, 22*, 281-7.

—— (1980) 'El concepto de región' in Asociación Geógrafos Españoles, *La región y la Geografía española*, Valladolid, 13-30.

—— (1981) 'Perspectivas de la Geografía en España y Portugal: Enseñanza, investigación y problemas profesionales' in *I Coloquio Ibérico de Geografía*, Universidad de Salamanca, 15-22.

—— (1983) 'La Geografía ibérica: Tendencias, resultados y problemas' in *II Col. Ibér. Geogr., Lisboa*.

13 CONCLUSION
P. Claval

At the end of the Second World War the era of national schools of geography was not yet over. In Germany, for example, the emphasis was on geomorphology and settlement, with an historical approach; the co-ordinating theme was landscape, and much time was spent in micro-scale studies, some in Germany but most abroad. In France, on the other hand, the regional theme dominated. The *doctorat d'État*, which opened the academic career, was traditionally a monograph on either a part of France or some section of the French Empire. The French emphasis was on evolution, on ways of life and on territorial organisation. And in the United States there was not one dominant school but two. Mid-west geographers started with the land-use map and minute economic observations; their main theoretical inspiration was from economics and social ecology and they stressed the present. At Berkeley, Carl Sauer had developed a very different discipline; as in Germany the emphasis was on landscape features, but the connection with cultural anthropology was stronger and the historical dimension less conspicuous.

Elsewhere, the inspiration was rather derivative. Britain had a tradition of exploration and of global geography (after Mackinder), but the majority of geographers were working in regional themes along French lines. French influence was strong in Spain, Portugal and Latin America, too, whereas in Italy and Romania, French and German themes intermingled. In Central and Eastern Europe, the main stimuli were from Germany, as was true also in Scandinavia — though Scandinavian geographers were well aware of activities in Great Britain, France and the United States. In the Netherlands, initial development was on French lines, but with a stronger emphasis on social problems. Finally, Russian geography — prominent around the turn of the century — was developing, after the 1917 revolution, a new type of economic analysis (pioneered by Baransky), but the majority of geographers worked on physical geography topics and their focus on landscape indicated a strong German influence.

The international scene in 1945 was dominated by the French, German and American traditions, with few feedbacks from other countries. The British situation was a peculiar one; the geographers there did not develop original conceptions but were able, through the imperial links, to be very influential on the development of the subject in the Dominions, in Africa, in the Indian subcontinent and in the Middle East.

In some small countries, original conceptions had developed as a result of individual travels during the formative years of academic careers. In addition, geographers there tended to be more open to the influence of other social and physical sciences. In the Netherlands, for example, there was a strong emphasis in social problems, as there was also in some Baltic universities, illustrated by the work of Edgar Kant; such orientations had little influence on geographers elsewhere before 1945.

One part of the German tradition was very specific to that country – the emphasis on research abroad. This dates from the work of the great pioneers - Humboldt and Richthofen – and was more conspicuous in the 1920s and 1930s than ever before. There was some spillover of this emphasis in foreign field-work in other countries. In France, for example, it was considered important to prepare doctoral dissertations on overseas topics, but for financial and career reasons only a few talented geographers were able to fulfil this prior to 1945.

Direct intercourse among geographers from various countries was limited to a few occasions, such as international congresses. The Amsterdam International Geographical Congress (IGC) in 1938, for example, was instrumental in the spread of new ideas – such as central-place theory – but in general each national school was self-contained and developed its own particular characteristics.

Until 1945 practically all geographical research was conducted by university academics. Departments were in most cases small, giving few possibilities for specialisation. Most of the students were destined to become school teachers, and only in a few countries (notably the Netherlands) was planning a significant career opening for geographers.

Growth and Internationalisation

Today the situation is very different from that just described. The discipline has grown spectacularly, built on a major increase in student numbers. Even in countries with long scientific traditions – such as France and Germany – the number of university teachers has increased between five- and tenfold; elsewhere, the growth has been even greater.

In some countries, the exclusive hold of the universities over geographical research has been broken. In Great Britain, the United States, Germany and Scandinavia, universities remain the major research centres, but increasingly money comes from national agencies. In France, Spain and Brazil there are national laboratories, maintained to undertake work on a variety of projects. And in Eastern Europe, the separation of teaching and research is virtually complete; the universities undertake the former role and the main research programmes are conducted in the geographical sections of the Academies of Science. In general, geography has entered the realm of 'big science'; more resources, more finance, but less freedom in the choice of research themes and topics. In Eastern Europe (such as the DDR) this transformation is virtually complete, with practically all research is conducted along lines determined by the state and linked to planning problems.

Changes have also come about because of a shift in the market for trained geographers. The secondary school systems can no longer absorb all of the products of the university departments. In Scandinavia and West Germany, this problem was exacerbated by a shift from traditional teaching of history and geography to global social science, a switch which had occurred earlier in the United States and accounts for the discipline's lack of weight there. In France, the beginnings of a similar shift are now in evidence. Thus other career outlets have been sought. Planning provided many opportunities in various countries at different times – the Netherlands since the 1930s, Britain in the 1940s-60s and France in the late 1960s-70s. Syllabuses have been altered as a consequence, with a shift from regional to systematic themes, the emergence of a more theoretical approach, and the burgeoning growth of urban studies.

The pattern of international communication has changed markedly too. Air travel and cheaper fares have allowed longer journeys to be undertaken, so that academics can visit foreign universities and attend international congresses regularly and frequently. In the 1930s the transfer of ideas was mainly achieved via the written word; it is now rooted much more in personal contacts. These developed first in the English-speaking world, leading to a large common store of geographical knowledge. Geographers in countries with a long tradition of English usage, such as the Netherlands and Sweden, adapted quite readily to this new situation, but for others it made for difficulties, especially the French and Germans who were losing their previous dominant positions.

Prior to 1945 there was one-way communication from the French, the German and the American schools to the rest of the world, with little feedback. Today the system is simpler: one-way communication from the English-speaking countries to the rest of the world. Because the ability to use other languages has declined drastically in the English-speaking world, international communication elsewhere depends increasingly on those geographers with contacts with English-speakers, and who can read and lecture in English.

The new patterns of communication have had other consequences. Before 1945 geography was highly compartmentalised in the universities. Each professor, with tenure for life, could establish his or her own orientation, free from any modes, fads and band-wagons. Traditions were thus maintained, and there was a healthy pluralism in the discipline. Where the universities are still organised along such hierarchical lines, as in Germany, this situation remains – accounting for the persistence of certain orientations there and the marginal status of newer developments. Almost everywhere else, however, the tempo of change has been fast. The market for young geographers in the universities is highly competitive, and for success individuals should be aware of developments in the subject. Thus there is a strong inclination to accept the newest set of ideas, whatever their intrinsic value. The result is either uniformity of geographical practice in many countries, as with the 'new' quantitative/theoretical geography of the English-speaking world in the 1960s, or a jungle of parallel claims to prominence, as in the 1970s.

The history of geographical thought in the 1960s and 1970s reflects new international communication patterns, therefore. It seemed then that traditional forms of geographical inquiry were being suppressed. Now it seems that they were being repressed but not destroyed, and the expressed unanimity of the 1960s has disappeared. Some have been revived since, suggesting that the university and research systems are sometimes open to the wrongs of intellectual terrorism.

Physical and Human

The period 1910-50 was characterised by the three major national schools, French, German and American, whose separate orientations reflected the problems of the discipline at that time. Geographers identified their subject with the natural sciences, but they were dealing mainly with social facts. There was unity in physical geography, based on the same set of theories, but not in human geography. The latter lacked a coherent theoretical framework (the nearest was drawn from ecology). Instead, human geographers followed one of three routes: the study of regional differentiation; the study of people-milieu relationships; and the study of landscapes. Interpretation could be either historical or functionalist. The three routes and two interpretative modes could be combined in a variety of ways, providing the basis for a separate 'school of thought'. There were fewer combinations than nations, however, and most of the viable ones were represented in the three main schools – French, German and American.

During the last 30 years much effort has been expended on the construction of a more coherent discipline. The unity of physical and human geography was seen as more apparent than real, with few logical roots – except in ecology. As a result, the two seem more independent now than a generation ago, especially where physical geographers have developed specialisations linking them closely with botanists, geologists, pedologists and meteorologists, whereas human geographers spend all of their time with economic and social models.

There is a possibility of a *rapprochement*, however, focused on ecology. It implies for physical geographers that they should make integrated studies of environments, whereas from human geographers a fresh look at the ecological basis of social life is required. Work along such lines is being prosecuted in several places: in Eastern Europe, because of the traditional interest in the physical components of landscapes and the popularity of analysing systems; in the United States, with the perceived need for an environmental grounding to geographical work – after several decades when the physical side of the discipline was virtually absent from university departments; and to a lesser extent in France and Great Britain.

In human geography, the developmental sequence has been the same nearly everywhere, apart from differences in timing – with the United States, Sweden and Great Britain slightly in the lead. The idea of a scientific revolution was widely accepted, with the emphases on quantitative methods and theory-building. Disillusion soon set in, however, and by the early 1970s new perspectives were being sought – phenomenological, humanist, radical, etc. Were these new scientific revolutions? Kuhn's views on the history of scientific thought were popular among geographers. But there were many mismatches between model and reality; there were too many breaks in a short time, and parallel developments rather than successive displacements were the general trend.

Other interpretations of the recent history of human geography are now being sought. In the 1980s few geographers consider themselves as natural scientists. Instead, they are increasingly conscious of links with the social sciences and the humanities, and they seek accounts of human distributions not only through environmental controls but also through the constraints of social life and relations. In the 1960s spatial models built on the concept of distance-decay provided the foundations for the search, and the positivist philosophy provided a popular approach to understanding. But the explanations available from this combination of positivism and spatial economics cover only a limited range of human distributions. Other models were sought, and geographers along with other social scientists discovered the significance of hermeneutics. Energy was channelled into the exploration of social and political models.

Homo oeconomicus is always and everywhere the same; he has no age, sex or religious experience. As long as geographers were content with economic reasoning, then a focus on economic organisation, and the related political and social institutions, was sufficient. There was no need to investigate human differences, idiosyncracies and ideologies. But now, the shifting emphasis demands such investigations, for the ways in which institutions work reflect the feelings of the people involved, while efficiency in communication reflects the level of confidence between the partners, itself an outcome of shared values and attitudes. So, as in the other social sciences, in the 1970s discussion in human geography focused on alternative (phenomenological, Marxist, idealist, etc.) models of man.

As a result of these developments, the coherence of human geography is greater than it was 20 years ago, and the links with other social sciences are stronger. The pervading atmosphere of disillusion and the sense of an impending crisis are less justified than they were in the 1950s. But they are there, and they are real, especially in North-west Europe and in the English-speaking world.

Structures and Movements

Before 1950 differences in the practice of geography reflected a diversity of conceptions of the discipline. In the 1980s easier international communications and a keen interest in theory-building have eradicated the major national differences. There is still a relative emphasis on regions in France, on landscapes and geomorphology in Germany, and on social aspects of distributions in the Netherlands, but these do not preclude consensus on what geography is and what are the best ways of doing it.

There are still national differences, however, and not of a residual character only. Many are new, with different origins from the old. Some are rooted in political systems and institutions, for example. Eastern European countries have developed strong research centres sponsored by the Academies of Science; they emphasise environmental studies, with reference to planning problems, and they try to develop an economic geography rooted in Marxism – with little apparent success; the methods they use are increasingly similar to those employed in the West.

The main oppositions, nevertheless, are among the Western countries themselves. The orientation of work is not now very dependent on levels of development (the same methods are being experimented with in Latin America as in North America), or on previous cultural traditions (Japanese geography followed an original course between 1880 and 1940, but does not do so now). The significant differences are in anxieties about the discipline, which are strongest in North-west Europe and North America. There are radical movements everywhere, it is true. They originated in North America and Great Britain. They are relatively superficial in continental and Southern Europe, expressed more through traditional forms of social unrest than on new critiques of the social order – anarchism in Spain, for example, and Marxism of the communist party brand in Italy, France and Latin America. The French radical movement is particularly weak, with left-wing ideologies losing their appeal to the intelligentsia after the Stalinist era.

The radicalism evident in the English-speaking world, and its Dutch, German, and Scandinavian equivalents, reflects dissatisfaction with the state of the world. It is not peculiar to geography, for similar attitudes are apparent in anthropology, economics and sociology. Radical geographers are critical of societies dominated by utilitarianism; their criticism of geography reflects its inability to disentangle what is a natural fact from that which is a human creation. Thus the new dividing line within the discipline is concerned less with disagreements about the ultimate goal of geographical work and more with disagreements about the nature of the world; the conflict is more ideological than strictly scientific.

What Next?

Cautious predictions about forthcoming transformations of the discipline are possible. In all of the developed countries, Western and Eastern, the era of easy, rapid growth is over. Everywhere there are restrictions in funding, less liberal science policies, and ageing research and teaching staffs. The possibilities of radical changes are few; to launch something new, it is usually preferable to appoint a young person than to ask an old professor to alter ideas, methods and interests.

Opposition between Third World and developed countries will almost certainly grow. The former will be less dependent on expatriates and higher degrees from overseas countries — with the result that diversity of geographical orientations will increase, to reflect the variety of cultural conditions.

And what of the radical movement; will it become dominant? Not as long as it relies on reductionist assumptions. But with a stronger theoretical base, including a better assessment of what makes people, institutions and organisations different, it could constitute the next step in the evolution of geography from a naturalist conception, to a social science, and then to a humanistic discipline.

Milton Keynes UK
Ingram Content Group UK Ltd.
UKHW031145141024
449569UK00024B/1068

9 781138 975125